Neoliberal Bio-Economies?

Kean Birch

Neoliberal Bio-Economies?

The Co-Construction of Markets and Natures

Kean Birch
York University
Toronto, ON, Canada

ISBN 978-3-319-91423-7   ISBN 978-3-319-91424-4   (eBook)
https://doi.org/10.1007/978-3-319-91424-4

Library of Congress Control Number: 2018944439

This Palgrave Macmillan imprint is published by the registered company Springer International
Publishing AG part of Springer Nature.
The registered company address is: Gewerbestrasse 11, 6330 Cham, Switzerland

For Sheila, Maple, and 'Pipsy'

ACKNOWLEDGMENTS

The research in this book was supported by an Insight Development Grant from the Social Sciences and Humanities Research Council (SSHRC) of Canada (Reference: 430-2013-000751).

I wish to thank the following people for their intellectual contribution to my ideas in this book, although they are in no way responsible for what I have produced: Jenn Baka, Kirby Calvert, Peter Kedron, Teis Hansen, Warren Mabee, Stefano Ponte, and Mark Winskel.

I also wish to thank the following people: the editorial and production team at Palgrave Macmillan, especially Rachel Krause Daniel and Kyra Saniewski, Emily Simmonds and Venilla Rajaguru for their research assistance, and all the interviewees who contributed to the project.

Parts of this book draw on previous research published under open access licenses: Chapter 4 is based on Birch, K. (2016) Emergent policy imaginaries and fragmented policy frameworks in the Canadian bio-economy, *Sustainability* 8(10): 1–16; and Chap. 5 is based on Birch, K. and Calvert, K. (2015) Rethinking 'drop-in' biofuels: On the political materialities of bioenergy, *Science and Technology Studies* 28 (1): 52–72.

As always, I owe most to Sheila and Maple—all three of us are expectantly awaiting the imminent arrival of 'Pipsy' who held on long enough for me to finish this book (almost) to deadline.

CONTENTS

LIST OF FIGURES

LIST OF TABLES

CHAPTER 1

Introduction

"We're the middle children of history, man. No purpose or place. We
have no Great War. No Great Depression. Our Great War's a spiritual
war... our Great Depression is our lives".
Tyler Durden, Fight Club *(1999)*

INTRODUCTION

Tyler Durden is wrong. We might expect as much from the imaginary
alter ego of the main character 'Jack' in the 1999 film *Fight Club*, but it's
worth emphasizing right off the bat. Our purpose, at its base, is to solve
the self-annihilation of humankind, and to do so sometime in the next
20–30 years. Without action now, anthropogenic climate change will erase
and erode our world as we know it, turning societies upside down, leading
countries to war, disrupting economies and markets, and destroying lives
and livelihoods. And, we most definitely have a fight on our hands to stop
it. Numerous businesspeople, politicians, commentators, and thinkers
stare humanity's possible extinction in the face with nary a blink; examples
abound and range from politicians like the US President Donald Trump
through the now infamous Koch brothers and other corporate managers
all the way to 'Global Thinkers' like Bjorn Lomborg. None of this is new
or news, it must be stressed, in that turning a blind eye to climate change
has a long, if not distinguished, history (Oreskes and Conway 2010).

Fortunately, we have options—many of them. Our fate is not set, our
history is not made—at least, not yet—but time is fast running out for all

© The Author(s) 2019
K. Birch, *Neoliberal Bio-Economies?*,
https://doi.org/10.1007/978-3-319-91424-4_1

of us on this planet to find ways to ameliorate the worst effects of climate change and adapt to the others. My aim in this book is to consider one such option or pathway to a low-carbon future, namely, the 'bio-economy'. Generally, the term bio-economy is used to describe an economy underpinned by the use of biological material like plants, residues, and waste as the input into the production of energy (e.g. biofuels), consumer goods (e.g. bioplastics), and manufacturing inputs (e.g. biochemicals), thereby replacing or substituting fossil fuels like oil, natural gas, and coal (Matthews 2009; Kircher 2012; El-Chichakli et al. 2016). As with most things in life, it is more complex than this short definition suggests, but I'll come back to the complicating factors in the following chapters. For now, it's easiest to think of the bio-economy as a wholesale shift in the way our economies—and necessarily our societies and polities—are organized and coordinated such that they are no longer based (primarily) on fossil fuels.

At present, the bio-economy is a rather high-level policy or expert concept that most publics have simply never heard about. Even though both the European Union (EU) and The White House produced bio-economy policy strategies in 2012 (CEC 2012; White House 2012), it is unlikely these were even mentioned in the mainstream media at the time. This lack of public awareness might be changing as more and more policy and social actors get behind the bio-economy—see, for example, these recent posters by the Ontario Farmers Association and Canadian Renewable Fuels Association (see Fig. 1.1)—or as activist groups criticize it (e.g. TNI 2015). However, these examples aside, it's likely that the bio-economy will remain a high-level policy vision and framework for the foreseeable future. That doesn't mean that the bio-economy is not influential; increasingly, it is emerging as a key policy vision and framework around the world as more and more countries develop their own policy strategies (Staffas et al. 2013; German Bioeconomy Council 2015; Birch 2016a).

Hopefully this book will contribute to creating greater awareness and understanding about the bio-economy, not only its potential as a solution to climate change but also its limits and how we might resolve those limits in the pursuit of a low-carbon and sustainable future. In this introduction to the book, I start by outlining the threats posed by climate change and the need to transition to a low-carbon future, before I briefly outline the analytical approach I take in understanding the bio-economy and its positive and negative implications. In particular, I stress the need to analyse the 'political-economic materialities' of the bio-economy, something I've written about elsewhere (Birch and Calvert 2015; Birch 2017), in order to understand how the co-construction of markets and natures underpinning the bio-economy affects policy strategies and industrial sectors.

Fig. 1.1 Posters by the Ontario Farmers Association (left) and Canadian Renewable Fuels Association (right). (Credit: Pictures taken by author)

OUR GLOBAL CLIMATE CHALLENGE

I started my research for this book in 2013 in order to understand the potential of different social and technical—or *socio-technical*—pathways as ways to ameliorate the worst, impending effects of climate change. The year before that I had read an article by the environmental campaigner Bill McKibben in *Rolling Stone* magazine. He highlighted three 'scary' numbers relevant to our current climate context:

- 2 °C
- 565 gigatons
- 2795 gigatons

McKibben points out that if we want to limit global temperature rises to 2 °C by the year 2100—the generally accepted *and* doable target agreed upon by diverse governments, agencies, and organizations around the world (e.g. HM Treasury 2006; Stern 2010)—then we cannot put more than 565 gigatons of CO_2 into the atmosphere. However, two major problems face us with this limit. First, at current rates, we will reach that limit by 2030. Second, and where the largest number comes in, the current fossil fuel reserves of companies (e.g. Exxon, Shell, Aramco) and governments (e.g. Saudi Arabia, Norway, Canada) around the world represent 2795 gigatons of CO_2. The real difficulty is that while:

> ...this coal and gas and oil is still technically in the soil...it's already economically aboveground – it's figured into share prices, companies are borrowing money against it, nations are basing their budgets on the presumed returns from their patrimony. (McKibben 2012)

This means that these fossil fuels are already effectively spent; we're already going to release five times the limit necessary to meet the 2 °C target as the result of current political-economic decisions and choices. The only way to stop this happening is to write off $20 trillion worth of assets that provide the resources for jobs, government spending, pensions and savings, and so on (also see Berners-Lee and Clark 2013; McKibben 2016). It is, therefore, vital to find alternative ways to not only organize and coordinate our economies but also to provide outlets for the massive capital investments currently sitting in coal beds, oil wells, and natural gas reservoirs. I'll come back to this energy-finance transition issue at several points later in this book.

Obviously, numerous academics, activists, environmental organizations, policy-makers, politicians, and others had been concerned with climate change well before my own interest in the topic. There have been many policy responses over the last three decades, stretching back at least to the 1992 *Rio Earth Summit* organized by the United Nations (UN). The Rio Summit involved the signing of the UN Framework Convention on Climate Change (UNFCCC), which was finally ratified in 1994, with the objective:

> ...to achieve, in accordance with the relevant provisions of the Convention, stabilization of greenhouse gas concentrations in the atmosphere at a level that would prevent dangerous anthropogenic interference with the climate system. Such a level should be achieved within a time-frame sufficient to allow ecosystems to adapt naturally to climate change, to ensure that food production is not threatened and to enable economic development to proceed in a sustainable manner. Article 2, UNFCCC 1992

Subsequently, the UNFCCC was extended in 1997 through the adoption of the Kyoto Protocol, which committed signatory countries to reducing their greenhouse gas (GHG) emissions, primarily resulting from fossil fuel energy production and consumption. In the first commitment period (2008–2012), this reduction was an average of five percent below

1990 levels. In the current commitment period (2013–2020), however, signatory countries are supposed to reduce their emissions to 18 percent below their 1990 levels by 2020. According to Oliver Tickell (2009: 8), the Kyoto Protocol proposed a series of primarily 'market' mechanisms to "achieve global climate neutrality" and "move towards an equitable low-carbon economy". Subsequent policy developments, at the 2007 Bali UNFCCC Conference of the Parties (COP13), for example, sought to identify future discussion topics and establish financing and technological support for the Global South (ibid.).

These policy responses have been—and continue to be with both COP15 (Copenhagen 2009) and COP21 (Paris 2015)—driven by 'market-based' approaches. Early examples include the EU's 2005 emissions trading system (ETS) (Lohmann 2010; Felli 2014), the Kyoto Protocol's Clean Development Mechanism (CDM) and Joint Implementation projects (Tickell 2009), and the emerging interest in so-called ecosystem services (Dempsey and Robertson 2012). Such policy responses are framed by specific techno-economic assumptions regarding the implications of climate change to our societies, exemplified by Nicholas Stern's famous report on the economic impacts of climate change (HM Treasury 2006). From this market-based perspective, environmental impacts are framed in terms of their political-economic costs and benefits, where these are calculated and accounted in monetary terms (e.g. market value). As such, they reflect what many scholars, and increasingly politicians, policy-makers, and journalists, call a *neoliberal* mindset (Birch 2017). As such, energy and GHG emissions have become a key battleground for wider debates about the political-economic future of our societies and economies (e.g. Huber 2013; Moore 2015).

<h2 style="text-align:center">LOW-CARBON TRANSITIONS: THE CASE OF THE BIO-ECONOMY</h2>

It is in this context that the bio-economy first emerges as an important policy vision and strategy for promoting alternative and (supposedly) sustainable energy (and other materials), specifically in the development of liquid biofuels for transport. Liquid biofuels are energy produced from biological matter, including food crops like sugar or corn, high-energy crops like canola, and non-food crops like *Miscanthus*, switchgrass, and other sources (Worldwatch Institute 2007). These liquid biofuels represent a growing proportion of bioenergy use around the world, although they are currently still dwarfed by the use of solid biofuels (e.g. wood, pellets, charcoal, etc.). The policy (and political) attraction of liquid bio-fuels (from now on biofuels) is primarily because they represent one of the few ways to replace oil-derived petroleum fuels used in cars, heavy goods vehicles, airplanes, and suchlike. Long distance travel will likely remain dependent on high-energy fuels for the foreseeable future, meaning that electrification will only go so far (e.g. municipal public transit).

Biofuels policy initially focused on 'first-generation' biofuels—essentially ethanol—produced from sugar, corn, and other high-starch crops (Mabee 2007). Examples of the expansion of these biofuels include the EU's 2003 *Biofuels Directive* (2003/30/E)—mandating that 5.75 percent of transport fuels be biofuels by 2010—and the USA's 2005 *Energy Policy Act*, which mandated 7.5 billion gallons of biofuels by 2012 (Worldwatch Institute 2007). Up until the mid-2000s, politicians, policy-makers, environmentalists, business, and international organizations had formed a coalition of sorts supporting the market extension of these bio-fuels as a sustainable alternative to fossil fuels (Mol 2007). However, by the mid-2000s, the love affair with biofuels ended as their economic, social, and environmental benefits became increasingly contested and con-testable (Smith 2010; Mohr and Raman 2013). It is in this context that I began my own interest in the bio-economy and the potential of conventional *and* especially advanced biofuels as a socio-technical pathway to a low-carbon future.

Societies change all the time as the result of revolutions in *socio-technical systems*. In thinking through these sorts of revolutionary transformations, the historian of technology Thomas Hughes (1983) sought to understand how particular scientific and technological artefacts (e.g. electric light-bulbs) are necessarily bound up with wider social, economic, and political changes in societal infrastructures and institutions (e.g. electricity genera-tion and distribution, energy markets, etc.). Another apt example is the rise to dominance of the automobile in the early-twentieth century and its influence in (re)configuring the material landscape of cities (e.g. suburbs), travel and mobility (e.g. highways), energy production and use (e.g. oil), and even culture (e.g. road movies) (Mitchell 2011; Huber 2013). How alternative energy sources like biofuels challenge, integrate, or diversify this fossil fuel-based socio-technical system is a key question to ask, and this book represents my attempt to address that issue.

Often presented in Schumpeterian terms as a process of 'creative destruction' (Mazzucato 2013), the wholesale transformation of socio-technical systems has given rise to a new scholarly field concerned with examining the *socio-technical transitions* from one 'regime' to another (Geels 2002; Geels and Schot 2007). From this perspective, a system comprises "a cluster of elements, including technology, regulations, user practices and markets, cultural meanings, infrastructure, mainte-nance networks, and supply networks" (Geels 2005: 681). As such, transitions have to be understood in terms of multi-scalar, multi-actor dynamic processes in which certain regimes (e.g. petroleum-based cars) have to be actively stabilized in the face of emerging and competing niche innovations (e.g. electric cars) that can enter, align, and then alter existing regimes. External pressures can also lead to change, in that broader societal pressures (e.g. global climate policy) can lead to changes in regimes (Geels 2002). Others have adopted and adapted this theoretical framework; for example, David Tyfield (2014: 586) notes its strength, primarily that the "focus on new technologies" is not "to the exclusion of both the irreducible social factors and the systemic nature of stabilised socio-technical settlements and their transition". He also highlights its weaknesses in theorizing 'power', namely, that "'power' enters the picture only to 'change' a system already there and conceptu-alized as stable" and it's often seen as a "nefarious force responsible for lock-in to dysfunctional systems" (Tyfield 2017: 3). There's much to admire, then, but also a need to push these debates further and from other perspectives.

In light of this debate, it's worth examining whether the bio-economy represents a socio-technical system that challenges the existing, dominant fossil fuel system, as well as examine the extent to which it can be pro-moted, supported, and rolled out through deliberate policy and political action as a response to worsening climate change effects. And the reverse can be examined too, the extent to which specific visions of the bio-economy and their enactment in (market development) policies can be resisted and challenged. Despite attempts otherwise, including my own (e.g. Birch 2016a), it's difficult to find any agreement on definitions of the bio-econ-omy, whether that is as a simple descriptive term or as an analytical con-cept. This will be evident throughout the book; the bio-economy is many things to many people (Frow et al. 2009).

Environmental Economic Geographies: Neoliberal Natures? Neoliberal Bio-Economies?

One particular issue that raises its head in relation to the notion of the bio-economy as a socio-technical transition is that the geographical dimensions of such transitions are often ignored or overlooked in existing research (Coenen et al. 2012; Truffer and Coenen 2012; Hansen and Coenen 2015; Calvert et al. 2017). For example, Truffer and Coenen (2012: 10) claim that the socio-technical transitions literature tends to conceptualize things like institutions (e.g. markets, regulation) in the abstract, whereas they argue it is necessary to consider the "interdependencies among institutional configurations in specific places". Similar arguments are made by Hansen and Coenen (2015) who note the importance of understanding the geographical specificity of inter-organizational relations to transitions. Of particular importance in this regard, the discussion of 'sustainability' transitions (e.g. to a bio-economy) necessitates a consideration of the wholesale disruption and derailment of prevailing, dominant regimes (e.g. fossil fuels) across an array of geographically specific infrastructures (e.g. distribution), institutions (e.g. oil markets), value chains (e.g. sourcing), and social values (e.g. driving) (Calvert et al. 2017). It is not, then, only a social and technical transformation, it is, as much, a *material* transformation, changing the social, technical, *and* material elements that make up specific systems (Birch 2013, 2016b, 2017; also Becker et al. 2016).

My approach in this book is driven by this understanding of system innovation. Previously, I've highlighted the limited engagement with materiality in the socio-technical transitions literature (Birch 2016b). Others have done the same; for example, Becker et al. (2016: 94) argue that the existing literature is often "devoid of scalar or spatial sensitivity". In relation to the bio-economy, I and others stress the need to analyse the interplay between political-economic and environmental processes in any attempt to understand the potential of the bio-economy to contribute to any transition to a low-carbon future (Birch and Calvert 2015; Calvert et al. 2017). In this book, I want to stress the importance of such *political-economic materialities* to the analysis of the bio-economy, especially for understanding how markets and natures are co-constructed, or evolve together. Theoretically, I follow the likes of Timothy Mitchell (2009, 2011) and others (e.g. Graham and Marvin 2001; Bakker and Bridge 2006; Lawhon and Murphy 2012; Becker et al. 2016) in bringing together materiality and socio-technical relations as a way to understand how biophysical materialities are necessarily political-economic, and vice versa. That is, how markets and natures are co-constructed.

These efforts represent my contribution to two emerging sub-fields in human geography: first, to *environmental economic geography* (EEG) which attempts to move away from treating nature and the environment as stable, uncontested categories *and* as "passive location, condition or resource factor input" (Heidkamp 2008: 62). A burgeoning area of interest, EEG has grown out of work on ecological modernization and considers the discursive and material constructions of nature, the diverse institutional fixes used to resolve environmental issues, the proliferation of stakeholders, the need for whole life-cycle analyses, and the co-constitutive nature of material and economic processes (e.g. Gibbs 1996, 2006; Bridge 2008; Hayter 2008; Soyez and Schulz 2008; Patchell and Hayter 2013; Kedron 2015). Second, to *neoliberal natures* which draws on critical political ecology in order to analyse the deployment of market-based mechanisms, processes, and institutions in the management of environmental problems (see Castree 2008a, b). Similar to EEG, this neoliberal natures literature highlights the recalcitrance of 'nature' and the need to understand how political economy and nature are co-constructed, both analytically and empirically.

Bringing these two sets of ideas to bear on the bio-economy as a socio-technical transition highlights how the bio-economy has to be understood in terms of the co-construction of markets and natures. This is necessary for at least two reasons. First, it goes beyond simplistic notions of neoliberalism as the deployment of market mechanisms to solve environmental problems; that is, the notion of a two-way constitutive process involves a reconfiguration of markets *and* economic processes. Second, it goes beyond notions of the bio-economy as only or mainly a policy vision and framework, which I've previously argued (e.g. Birch et al. 2010; Levidow et al. 2012, 2013), while acknowledging that such discursive imaginaries still have particular, constitutive effects (see Hilgartner 2007, 2015; Birch et al. 2014). This critical discourse analysis is helpful up to a certain point, but it becomes increasingly necessarily to go beyond such an approach when considering the manifestation of the bio-economy in production processes and standards, supply chains and infrastructures, and consumer or commercial products like biofuels.

Empirical Material and Outline of the Book

In this book, my empirical focus is the development of advanced biofuels in Canada. Although liquid biofuels have been around for some time, going back at least to the early-twentieth century, they have emerged as a pressing global issue in the last decade or so. A range of societal challenges—such as energy insecurity, weak rural development, and climate change—at the start of the twenty-first century have provided the rationale for a strong policy push behind conventional and advanced biofuels. And this has happened around the world. However, problems with conventional biofuels—of which the food versus fuel debate has become emblematic—have meant that advanced biofuels are increasingly in the spotlight as a potential technological pathway towards a low-carbon future. For my purposes, then, biofuels represent a useful case study of the emergence of a bio-economy: they are a relatively recent policy agenda; they are characterized as a socio-technical solution to societal challenges; they are derived from bio-based materials, rather than fossil fuels; and their promotion and success are still very much a contested and contestable proposition.

My analysis draws on a range of empirical materials, including primary and secondary data sources; in particular, I draw on the findings from 39 in-depth interviews with Canadian policy-makers, business people and trade associations, energy experts, and civil society organizations and associations (see Table 1.1). These interviews were undertaken in 2014 and 2015. All of these empirical materials come from a three-year project looking at the emergence and development of advanced biofuels in Canada. The project

Table 1.1   Research interviews

Interviewees	Number	Examples
Policy-makers	5	Federal ministries and agencies
		Provincial ministries and agencies
Business	24	Biofuel companies
		Trade associations (biotech, biofuels, feedstock suppliers)
		Consultancies
Energy experts	5	Academia
		Consultancies
		Biofuel companies
Civil society	5	Academia
		Non-profits

had three main objectives: (1) to analyse the policy visions, narratives, and priorities underpinning biofuels policy in Canada; (2) to examine the political-economic and ecological processes along advanced biofuel value chains in order to unpack how economic, environmental, and discursive processes configure this value chain; and (3) to identify the biophysical and energetic constraints on developing and commercializing advanced biofuels and whether these materialities disrupt existing infrastructures and institutions. In drawing on these interviews, my aim is to analyse the political-economic and environmental processes at play in the bio-economy, rather than dissect the meanings and values of these social actors. Consequently, I use the interview material here to explore social processes—such as imaginaries and policy frameworks—rather than the values and meanings ascribed to social phenomena by the informants themselves. I supplement this analysis by drawing on a range of secondary material (e.g. policy documents) produced by government, industry, and civil society.

I start the book with the theoretical framework in Chap. 2, where I outline in more depth the analytical approach I briefly sketched out in this introduction. In particular, I aim to synthesize insights from academic and, to a lesser extent, policy debates on the introduction and implications of market-based mechanisms in environmental policy (or 'neoliberal natures'), environmental economic geography (EEG), and work in science, technology, and innovation studies (STIS) that focuses on socio-technical transitions, especially in relation to the bio-economy. My specific analytical aim is to develop the concept of *political-economic materialities* as a way to understand societal transitions like the bio-economy. Following on from this, in Chap. 3, I provide some background to the 'emerging' bio-economy as a specific societal transition pathway. My intent in this chapter is to provide readers with a relatively detailed, although primarily descriptive, outline of the origins and development of the bio-economy as a policy agenda around the world. I outline the specific Canadian policy visions and strategies in Chap. 4, drawing on empirical material from a number of in-depth interviews with policy-makers, industry, and civil society. In this chapter, I analyse the discourses, narratives, and imaginaries that frame ('neoliberal') bio-economies in Canada, showing how particular policy visions are generative or performative through directing resources towards particular conceptions of the bio-economy.

In Chap. 5, I analyse how certain forms of bio-based energy, product, and input are enrolled in enabling and legitimating specific forms of biofuels. Of particular importance here are so-called 'drop-in' biofuels that are developed and configured specifically to be compatible with existing infrastructures, institutions, and value chains while ignoring a raft of other political, policy, and political-economic issues that they raise. Following on from this, in Chap. 6, my aim is to illustrate the material limits to policy visions and strategies visions, especially those driven by market-based notions of political-economic order. I examine the policy instruments and tools used to construct the bio-economy as a (neoliberal) market, including a range of 'market development' policies ranging from labelling, standards, subsidies, mandates, and so on. This sets up Chap. 7 which concerns the co-construction of markets and natures in the development of advanced biofuels in Canada. My intention in Chap. 7 is to show how the development of advanced biofuels is constituted by the interaction of economic and environmental processes, such that the expectations many policymakers, business people, and others have of the bio-economy and advanced biofuels require significant re-assessment. I finish the book in the "Conclusion" with a discussion of alternative bio-economies that could represent other potential pathways towards societal transition.

References

Bakker, K. and Bridge, G. (2006) Material worlds? Resource geographies and the 'matter of nature', *Progress in Human Geography* 30 (1): 5–27.

Becker, S., Moss, T. and Naumann, M. (2016) The importance of space: Towards a socio-material and political geography of energy transitions, in L. Gailing and T. Moss (eds) *Conceptualizing Germany's energy transition*, London: Palgrave Macmillan, pp. 93–108.

Berners-Lee, M. and Clark, D. (2013) *The burning question*, London: Profile Books.

Birch, K. (2013) The political economy of technoscience: An emerging research agenda, *Spontaneous Generations: A Journal for the History and Philosophy of Science* 7(1): 49–61.

Birch, K. (2016a) Emergent policy imaginaries and fragmented policy frameworks in the Canadian bio-economy, *Sustainability* 8(10): 1–16.

Birch, K. (2016b) Materiality and sustainability transitions: Integrating climate change into transport infrastructure in Ontario, Canada, *Prometheus: Critical Studies in Innovation* 34(3–4): 191–206.

Birch, K. (2017) *A research agenda for neoliberalism*, Cheltenham: Edward Elgar.

Birch, K. and Calvert, K. (2015) Rethinking 'drop-in' biofuels: On the political materialities of bioenergy, *Science and Technology Studies* 28: 52–72.

Birch, K., Levidow, L. and Papaioannou, T. (2010) *Sustainable capital?* The neo-liberalization of nature and knowledge in the European knowledge-based bio-economy, *Sustainability* 2(9): 2898–2918.

Birch, K., Levidow, L. and Papaioannou, T. (2014) Self-fulfilling prophecies of the European knowledge-based bio-economy: The discursive shaping of institutional and policy frameworks in the bio-pharmaceuticals sector, *Journal of the Knowledge Economy* 5 (1): 1–18.

Bridge, G. (2008) Environmental economic geography: A sympathetic critique, *Geoforum* 39: 76–81.

Calvert, K., Kedron, P., Baka, J. and Birch, K. (2017) Geographical perspectives on sociotechnical transitions and emerging bio-economies: Introduction to a special issue, *Technology Analysis & Strategic Management* 29(5): 477–485.

Castree, N. (2008a). Neoliberalising nature: The logics of deregulation and reregulation, *Environment and Planning A* 40: 131–152.

Castree, N. (2008b) Neoliberalising nature: Processes, effects, and evaluations, *Environment and Planning A* 40: 153–173.

CEC. (2012) *Innovating for sustainable growth: A bioeconomy for Europe* [COM(2012) 60 Final], Brussels: Commission of the European Communities.

Coenen, L., Benneworth, P. and Truffer, B. (2012) Towards a spatial perspective on sustainability transitions, *Research Policy* 41(6): 968–979.

Dempsey, J. and Robertson, M. (2012) Ecosystem services: Tensions, impurities, and points of engagement within neoliberalism, *Progress in Human Geography* 36(6): 758–779.

El-Chickakli, B., Braun, J., Lang, C., Barben, D. and Philp, J. (2016) Five corner-stones of a global bioeconomy, *Nature* 535: 221–223.

Felli, R. (2014) On climate rent, *Historical Materialism* 22(3–4): 251–280.

Frow, E., Ingram, D., Powell, W., Steer, D., Vogel, J. and Yearley, S. (2009) The politics of plants. *Food Security* 1(1), 17–23.

Geels, F. (2002) Technological transitions as evolutionary reconfiguration processes: A multi-level perspective and case study, *Research Policy* 31: 1257–1274.

Geels, F. (2005) The dynamics of transitions in socio-technical systems: A multi-level analysis of the transition pathway from horse-drawn carriages to automobiles (1860–1930), *Technology Analysis & Strategic Management* 17(4): 445–476.

Geels, F. and Schot, J. (2007) Typology of sociotechnical transition pathways, *Research Policy* 36(3): 399–417.

German Bioeconomy Council. (2015) *Bioeconomy policy: Synopsis and analysis of strategies in the G7*, Berlin: Office of the German Bioeconomy Council.

Gibbs, D. (1996) Integrating sustainable development and economic restructuring: A role for regulation theory? *Geoforum* 27(1): 1–10.

Gibbs, D. (2006) Prospects for an environmental economic geography: Linking ecological modernization and regulationist approaches, *Economic Geography* 82(2): 193–215.

Graham, S. and Marvin, S. (2001) *Splintering urbanism*, London: Routledge.

Hansen, T. and Coenen, L. (2015) The geography of sustainability transitions: Review, synthesis and reflections on an emergent research field, *Environmental Innovation and Societal Transitions* 17: 92–109.

Hayter, R. (2008) Environmental economic geography, *Geography Compass* 2: 1–20.

Heidkamp, P. (2008) A theoretical framework for a 'spatially conscious' economic analysis of environmental issues, *Geoforum* 39: 62–75.

Hilgartner, S. (2007) Making the bioeconomy measurable: Politics of an emerging anticipatory machinery, *BioSocieties* 2: 382–386.

Hilgartner, S. (2015) Capturing the imaginary: Vanguards, visions, and the synthetic biology revolution, in S. Hilgartner, C. Miller and R. Hagendijk (eds) *Science and democracy: Making knowledge and making power in the biosciences and beyond*, London: Routledge, pp. 33–55.

HM Treasury. (2006) *Stern review on the economics of climate change*, London: HM Treasury, available at: http://webarchive.nationalarchives.gov.uk/+/http://www.hm-treasury.gov.uk/sternreview_index.htm (accessed February 2018).

Huber, M. (2013) *Lifeblood*, Minneapolis: University of Minnesota Press.

Hughes, T. (1983) *Networks of power*, Baltimore: John Hopkins University Press.

Kedron, P. (2015) Environmental governance and shifts in Canadian biofuel production and innovation, *The Professional Geography* 67(3): 385–395.

Kircher, M. (2012) The transition to a bio-economy: National perspectives, *Biofuels, Bioproducts & Biorefining* 6(3): 240–245.

Lawhon, M. and Murphy, J. (2012) Socio-technical regimes and sustainability transitions: Insights from political ecology, *Progress in Human Geography* 36(3): 354–378.

Levidow, L., Birch, K. and Papaioannou, T. (2012) EU agri-innovation policy: Two contending visions of the knowledge-based bio-economy, *Critical Policy Studies* 6: 40–66.

Levidow, L., Birch, K. and Papaioannou, T. (2013) Divergent paradigms of European agro-food innovation: The knowledge-based bio-economy (KBBE) as an R&D agenda, *Science, Technology, and Human Values* 38: 94–125.

Lohmann, L. (2010) Neoliberalism and the calculable world: The rise of carbon trading, in K. Birch and V. Mykhnenko (eds) *The rise and fall of neoliberalism*, London: Zed Books, pp. 77–93.

Mabee, W. (2007) Policy options to support biofuel production, *Advances in Biochemical Engineering* 108: 329–357.

Matthews, J. (2009) From the petro-economy to the bioeconomy: Integrating bioenergy production with agricultural demands, *Biofuels, Bioproducts and Biorefining* 3: 613–632.

Mazzucato, M. (2013) *The entrepreneurial state*, London: Anthem Press.

McKibben, B. (2012) Global warming's terrifying new math, *Rolling Stone* 19 July, retrieved from: https://www.rollingstone.com/politics/news/global-warmings-terrifying-new-math-20120719.

McKibben, B. (2016) Recalculating the climate math, *New Republic* 22 September, retrieved from: https://newrepublic.com/article/136987/recalculating-climate-math.

Mitchell, T. (2009) Carbon democracy, *Economy and Society* 38 (3): 399–432.

Mitchell, T. (2011) *Carbon democracy*, London: Verso.

Mohr, A. and Raman, S. (2013) Lessons from first generation biofuels and implications for the sustainability appraisal of second generation biofuels, *Energy Policy* 63: 114–122.

Mol, A. (2007) Boundless biofuels? Between environmental sustainability and vulnerability, *Sociologia Ruralis* 47(4): 297–315.

Moore, J.W. (2015) *Capitalism in the web of life*, London: Verso.

Oreskes, N. and Conway, E. (2010) *Merchants of doubt*, New York: Bloomsbury.

Patchell, J. and Hayter, R. (2013) Environmental and evolutionary economic geography: Time for EEG²? *Geografiska Annaler B* 95(2): 111–130.

Smith, J. (2010) *Biofuels and the globalization of risk*, London: Zed Books.

Soyez, D. and Schulz, C. (2008) Editorial: Facets of an emerging environmental economic geography (EEG), *Geoforum* 39: 17–19.

Staffas, L., Gustavsson, M. and McCormick, K. (2013) Strategies and policies for the bioeconomy and bio-based economy: An analysis of official national approaches, *Sustainability* 5: 2751–2769.

Stern, N. (2010) *A blueprint for a safer planet*, London: Vintage Books.

Tickell, O. (2009) *Kyoto2*, London: Zed Books.

The White House. (2012) *National bioeconomy blueprint*, Washington DC: The White House.

TNI. (2015) *The bioeconomy: A primer*, Amsterdam: Transnational Institute.

Truffer, B. and Coenen, L. (2012) Environmental innovation and sustainability transitions in regional studies, *Regional Studies* 46(1): 1–21.

Tyfield, D. (2014) Putting the power in 'socio-technical regimes' – E-mobility transition in China as political process, *Mobilities* 9(4): 585–603.

Tyfield, D. (2017) *Liberalism 2.0 and the rise of China*, London: Routledge.

Worldwatch Institute. (2007) *Biofuels for transport*, London: Earthscan.

CHAPTER 2

Neoliberal Bio-Economies?

INTRODUCTION

What is the environment? What is nature? And how do we understand their relationship to political economy? Obviously, these are not new questions, and many others—ethicists, political ecologists, geographers, economists, and so on—have asked and continue to ask similar questions, as well as offer insightful analyses. Terms like 'anthropocene' or 'capitalocene', for example, are now commonly used to define the (relatively short) geological epoch dominated by humans (e.g. Moore 2015; Bonneuil and Fressoz 2017). That being said, there remains a tendency in political and policy circles—and perhaps amongst various publics—to assume that the environment and nature are something 'out there' that we can 'freely' extract resources from and then dump our waste into, sometimes with regard for the damage this does but often with little regard at all (Kenney-Lazar and Kay 2017). Our attitudes to 'nature'—as a shorthand for the biophysical processes and systems that sustain life on Earth—usually depend on where and with whom we live, how we are socialized, and the implications that caring about nature has for our livelihoods. In this book, I'm specifically focusing on attitudes to nature in the 'Atlantic heartland' of contemporary 'neoliberal' capitalism covering Canada and the USA in North America and the member states of the European Union (EU) (Jessop 2016).

© The Author(s) 2019
K. Birch, *Neoliberal Bio-Economies?*,
https://doi.org/10.1007/978-3-319-91424-4_2

It is perhaps notable that although we have known about the environmental problems caused by industrial capitalism since at least the nineteenth century, it was not until the mid-twentieth century that environmental movements arose to not only highlight the damage we do to the planet but also seek to reduce or ameliorate it. Today, 'sustainability' is a word most children hear before they leave school, although perhaps not with the same connotation in every country. Wherever we hear it, however, sustainability and environment are (almost) always inextricably linked to 'economy'—shorthand for the socio-political arrangement of production, distribution, consumption, and disposal of the materials used to live our lives. As a relatively recent word *and* concept, sustainability can be traced back to fears about the 'limits to growth' of over-using natural resources and the environmental, and the social problems this will cause (Club of Rome 1972). Its most popular incarnation is in the 1987 *Brundtland Report*'s definition of 'sustainable development' as economic development that "meets the needs of the present without compromising the ability of future generations to meet their needs" (World Commission of Environment and Development 1987: 41). As Birner (2018: 25) notes, this conception of 'sustainable' combines concerns about "environmental problems in industrialized countries" with the "poverty and population pressures" in the Global South. As such, nature and economy are not treated as separate and probably cannot be treated as such.

My intention is not to imply that concepts like sustainability are problematic *because* they position nature in relation to economy; rather, it is that *how* we connect nature and economy in our thinking matters. Take the bio-economy, for example. It's a policy and political concept that emerged within a particular political-economic way of understanding the world, often conceptualized as 'neoliberalism' (Birch 2006, 2017; Tyfield 2017). According to many thinkers, neoliberalism entails a specific worldview, one based on the *assumption*—and I call it that because neoliberalism is both an analytical and normative perspective—that markets are the best way to organize society and, by extension, our relationship to nature. Taking this neoliberal worldview to heart means treating nature as any other entity, process, or relationship, and letting the market resolve our environmental problems through the price mechanism—as well as poverty and population pressures. It follows from this that if people want to 'save' nature, then they will pay for pristine landscapes; if they do not, then they will not pay for it. Obviously, neoliberal perspectives only allow us to connect nature and economy in one way, as a question of price determined by market competition. As such, nature is treated as a simple calculation of the price people are willing to pay for their preference (e.g. love of hillwalking, wilderness, protecting other species, etc.). Nature is positioned as something out there we have the right to take and use as we see fit, as long as we are willing to pay for it. Arguably, this is how nature is treated in visions of the bio-economy; as a resource we can adapt and integrate as the underpinning biophysical material driver of our economic system. There are many other ways to understand the relationship between nature and economy, however, although my aim in this book is to examine the 'neoliberal' perspective in more depth using the bio-economy as an apposite example.

Understanding neoliberal approaches to nature helps to unpack current ways of resolving environmental problems, and any other problems this causes, as well as considering alternative approaches to resolving environmental problems. Does everything have to be treated as a market? My view is that it does not. However, I also think that much of the literature on so-called neoliberal natures (see Castree 2008a, b, 2010a, b; Bakker 2010), which provides the analytical starting point for many of my arguments, tends to treat 'the market' unproblematically as a political-economic process outside of nature, even an aberration of nature, while valorizing nature's resistance to it. In this chapter, in contrast, my aim is to examine how nature and economy are inextricably entwined, meaning that there is no 'neoliberalism' without 'nature'—in that markets are very much material entities—although in their hybridity it might be possible to argue that both come away from their entanglements as something else, something different from neoliberalism and from nature.

I start the chapter by outlining neoliberalism as a concept, as well as its origins and the major debates about its usefulness as an analytical category. I then examine the neoliberal natures literature, unpacking the importance of materialities as a way to problematize the idea that markets are imposed on nature. This leads me into a discussion of my main approach in this book, namely, *environmental economic geography*— broadly speaking. I finish by considering the importance of biophysical materialities to understanding the relationship between nature and economy, especially as this relates to innovation processes and systems (e.g. advanced biofuels) deployed to resolve environmental problems like climate change.

Understanding Neoliberalism

As a word, neoliberalism barely raises an eyebrow nowadays—at least amongst people of a certain political persuasion. Any left-leaning person in the Atlantic heartland has likely heard the term, whether or not they know what it means. For example, it is regularly used by writers, journalists, and commentators in media outlets like *The Guardian* newspaper (Dunn 2017). It has become so commonplace that it has even eclipsed 'free market' as a term to describe a particular policy stance and agenda based on market principles (Birch 2017: 62). While it's usually used in a pejorative sense, a quick search of contemporary social media reveals a growing adoption of the term as a positive moniker to differentiate right-leaning pro-globalization and pro-free trade supporters from the more rabid nativists behind the rise of far-right movements across Europe and North America (Bowman 2016). Not wishing to belabour the point any further, neoliberalism is a contested concept, and this is as much the case in academia as it is in public discourse (e.g. Venugopal 2015; Storper 2016; Birch 2017).

All that being said, I'm going to provide an analytical outline of neoliberalism here so that readers can understand where the later debates on neoliberal natures are coming from. At its base, neoliberalism is a concept used to define and analyse the insertion or installation of (free) markets as the primary organizing principle and institution of economic, social, and political life. A number of analytical approaches to understanding neoliberalism have emerged over the last decade or so (see Birch 2015, 2017), including:

- *Marxist* analyses by scholars like Dumenil and Levy (2004) and Harvey (2005) which theorize neoliberalism as a class-based project to restore elite class power through a variety of mechanisms (e.g. financialization, dispossession) that redirect income upwards towards the dominant political-economic classes. In this framework, though, markets and market competition simply end up as an ideological cover for whatever enables the elite class to accumulate the most income.
- *Foucauldian* analyses by scholars like Dardot and Laval (2014) and Brown (2015) which build on the work of Michael Foucault (2008) by arguing that neoliberalism represents a form of market-based 'governmentality'—or art of governance. As such, neoliberalism involves the production of new subjectivities that transform people

into 'entrepreneurs of the self'. How this happens, however, is not always clear nor are contemporary transformations necessarily dominated by market-based perspectives.

- *Epistemological* analyses by scholars like Mirowski (2013) and collaborators which focus explicitly on the epistemic assumptions and community that shape policy and politics. Whether these assumptions translate into 'uncorrupted' policy prescriptions is an issue worth considering, however, since many others note the hybridity of neoliberalism.
- *Processual* analyses primarily associated with human geographers, especially people like Peck (2010) and collaborators (e.g. Peck and Tickell 2002; Brenner et al. 2010), which tend to stress this very hybridity of neoliberalism—hence the conception of it as a process. As the primary approach to understanding neoliberal natures, I'm going to discuss these analyses shortly.

There are several other approaches to analysing neoliberalism, including ideational, institutional, and state-theoretic perspectives (Birch 2015), as well as chunky handbooks covering a diverse array of topics (e.g. Springer et al. 2016; Cahill et al. 2018). It's a topic and concept that continues to attract new ways of thinking and new case studies. Nevertheless, it's often difficult to pin down exactly what people mean when they use the concept of neoliberalism. I'd even argue that the very proliferation of ideas about and research on neoliberalism is part of the reason why it's so difficult to say what exactly we mean when we use the term.

With all this in mind, it's important to define the particular analytical conception of neoliberalism I'm using here; otherwise it's difficult to differentiate between processes (e.g. privatization) and their causes (e.g. government policy) and effects (e.g. market concentration). I'm drawing on the processual approach prevalent in human geography and most commonly associated with the work of economic geographers like Jamie Peck, Adam Tickell, Wendy Larner, and others. Rather than an outcome, effect, or condition, these scholars conceptualize neoliberalism as a process; that is, as *neoliberalization*. As such, neoliberalism cannot simply be understood as the withdrawal of the state, a too frequent definition, but rather as the "*mobilization of state power in the contradictory extension and reproduction of market(-like) rule*" (Tickell and Peck (2003: 166—emphasis in original). From this perspective, markets are not assumed to swoop in unhindered once the state withdraws—or is withdrawn—under deregulatory pressures. Instead, neoliberalization involves an "extensive deconstruction and reconstruction of institutions, often in the name or in the image of 'markets'" (p. 167–8), meaning that neoliberalization leads to "a tangled web of state-regulated oligopolies, profit-orientated enclaves and pseudo markets" (p. 167). As this would imply, neoliberalization is both a spatial *and* temporal process in that it unfolds over space and time in different and varied ways.

In their most well-read piece on neoliberalization, Peck and Tickell (2002) outline several phases of neoliberalization, primarily differentiating between *roll-back* and *roll-out* phases—as well as an earlier *proto*-phase. According to Peck and Tickell, the roll-back phase is characterized by a deregulatory focus associated with key politicians like Margaret Thatcher (UK) and Ronald Reagan (USA) in the 1980s and early 1990s. Here, neoliberalization primarily involved the mobilization of state power in the (often violent) restructuring of the "Keynesian-welfarist settlement" characteristic of the post-WW2 period (p. 388). Subsequent impacts of this deregulatory fervour—including rising unemployment, poverty, and inequality—engendered a new, roll-out phase of neoliberalization in which the Third Way governments of people like Bill Clinton (USA), Tony Blair (UK), and Gerhard Schroder (Germany) sought to transform government and the state through the introduction of markets into areas like public service delivery. As an institution- or state-building phase, roll-out neoliberalization lasted from the mid-1990s until the global financial crisis in 2007–2008 when many people began to speculating about the end of neoliberalism (e.g. Birch and Mykhnenko 2010). Since then, however, others like Sidaway and Hendrikse (2016) have suggested that we are merely entering a new phase, which they call 'neoliberalism 3.0', in which the state and market become even more intermeshed.

A diverse and varied array of processes (e.g. privatization, marketization, commodification), forces (e.g. state, capital), agents (e.g. business, international policy-makers, NGOs), discourses (e.g. new public management, competitiveness), and institutions (e.g. law, education, market) are implicated in neoliberalization, meaning that in every case neoliberalism has to be thought of as particular, as plural, as hybrid, or as variegated (Larner 2003; Castree 2006; Brenner et al. 2010; Peck 2010). For example, Birch and Siemiatycki (2016) outline a range of markets and marketization processes at play in the public-private restructuring of public infrastructure developments and public services delivery. Similarly, and as I discuss in the next section, many writers dealing with the neoliberalization of nature highlight the range of neoliberal processes at play in the transformation of nature into a market(able) entity. However, it's important to stress that this critical perspective of neoliberalism is also based on the notion that neoliberalization processes are always layered on top of existing institutions, creating place-specific forms of change (Brenner et al. 2010). Hence, it makes little sense to talk of 'neoliberalism' as a monolithic thing; instead, it is important to examine how particular places, sites, institutions, and communities are configured and reconfigured through the process of neoliberalization.

NEOLIBERALIZING NATURE

One critical site for examining this neoliberalization process is in reference to the environment and nature. A large literature on so-called neoliberal natures has built up over the last few years addressing issues ranging across agriculture (Roff 2008; Essex 2016), biofuels (Levidow et al. 2012), climate change (e.g. Bailey 2007a, b; Lohmann 2010), conservation (e.g. Büscher et al. 2014), ecosystem services (e.g. Dempsey and Robertson 2012), forestry (e.g. Prudham 2005), genetics and genomics (McAfee 2003), planetary environmental governance (Wainwright and Mann 2013), water (e.g. Loftus and Budd 2016), and much else besides. It'd be difficult to succinctly summarize all the work that has been done in this area, which isn't my intent here. Luckily, though, there are a number of detailed reviews of this literature that have been produced over the last few years (e.g. McCarthy and Prudham 2004; Castree 2008a, b, 2010a, b; Himley 2008; Bakker 2009, 2010; Collard et al. 2016). I mainly draw on these reviews in the following discussion.

According to this neoliberal natures literature, proponents of market-based instruments and mechanisms for (re)solving environmental issues, and their attendant socio-material problematics (e.g. population pressures, poverty, inequality, etc.), stress the benefits of a range of political-economic processes (e.g. privatization), policies (e.g. carbon trading), and technologies (e.g. biotechnology). Markets are meant to solve problems with the over-use of resources (through the extension of private ownership), externalities (through the creation of new commodities), and ecological and agricultural efficiencies (through new production technologies). Much of the critique of this market-based approach to environmental problems centres on the conception of neoliberalism as a process—that is, neoliberalization—which I've outlined above. These critiques are concerned with the specificities of the neoliberalization process, especially the transformations engendered by privatization, commodification, marketization, deregulation, and so on, as well as the particularities of the (natural) case or example at hand, whether that is environmental conservation, water quality, environmental impacts, climate change, or otherwise. I've tried to present some of these issues in Table 2.1 below, although this is obviously a simplification of a far more in-depth research programme. Table 2.1 contains a range of 'neoliberal' processes—by which I mean 'market-based'—ranging from privatization to depoliticization, as well as examples of how they relate to nature. For example, privatization involves the sale of public assets, like forests or woodlands, to private sector actors; it is different, although similar, to dispossession which involves the wholesale transfer of public assets to the public sector (e.g. appropriation of land to build a hydro dam).

Table 2.1 Neoliberal natures literature

Process	Transformation	Example
Privatization	Sale of public assets to private sector	Sale of forest
Marketization	Use of markets (proxies) and private sector in provision of public goods (e.g. outsourcing)	Building sustainable or climate-ready infrastructure
Commodification	Production of tradable good or service	Property rights over seeds
Commercialization	Valuation of existing asset or commodity in monetary terms	Notion of environmental services
Austerity	Introduction of fiscal cuts	State incapacity to enforce environmental regulations
Deregulation	Liberalization of regulatory rules	Changing trade rules and environmental protection
Reregulation (voluntary)	Setting of voluntary standards, codes, and principles	Industry-sponsored sustainability certification
Reregulation (mandatory)	Setting of new mandatory regulatory rules	Emissions trading
Governance devolution and rescaling	Rescaling of environmental regulations	Shifting responsibility for resources (e.g. forest) to local community
Dispossession	Transfer of public assets to private sector	Appropriation of land to build hydro dam
Depoliticization	Reducing government responsibility for citizen's life	Introduction of renewable energy feed-in tariffs

Source: Incorporating elements from McCarthy and Prudham (2004); Castree (2008a, 2010a, b); Bakker (2009, 2010)

As should be evident from Table 2.1, examples of the neoliberalization of nature are pretty varied and diverse. Moreover, some are more relevant to discussions of the bio-economy than others. A number of authors have focused, for example, on the implications of neoliberalization at the 'molecular scale' (McAfee 2003). Scholars like McAfee (2003) and Prudham (2007) have criticized the commodification of genetics and genomics through the use of intellectual property rights to enclose genetic material (e.g. cell lines) or genetic research (e.g. biomarkers). According to Prudham (2007), this commodification entails both a 'stretching' and 'deepening' of commodity production and circulation in order to open up new markets and/or turn more things into commodities (respectively), thereby marking new ways to integrate nature in the economy. However, as McAfee (2003) argues, forms of neoliberalization are necessarily based on particular—and often inaccurate—assumptions about and framings of nature, especially at the genomic scale where genetic determinism (e.g. one gene leading to a specific outcome) is built into neoliberal processes of commodification (cf. the complexities of ecosystem thinking) (also Essex 2008). Alongside these analyses of neoliberal misconceptions about biological materiality, other authors argue that neoliberalism and science can be compatible bedfellows (e.g. Gareau 2008; Birch et al. 2010), even leading to environmental benefits (Bakker 2009). It is important to remember, in this regard, that neoliberalism is not anathema to nature, or to our understandings of nature—and the same would likely apply to markets generally.

There are several aspects of this literature worth considering more critically, especially as these relate to the materialities of nature and markets. First, a significant element of the analytical worth of the neoliberal natures literature depends on the theoretical consistency across the various research projects and agendas about what precisely 'neoliberalization' applies to what 'nature' (Bakker 2009). As such, it is worth examining one process in more detail to unpack some of the inconsistencies that scholars have had to address. I focus on 'marketization' here because I'm interested in the co-construction of markets and nature in this book. In an early review, Castree (2008a: 142) defines marketization as "the assignment of prices to phenomena that were previously shielded from market exchange or for various reasons unpriced". In a later review, Castree (2010a, b: 1728) defines it as "rendering alienable and exchangeable things that might not previously have been subject to a market". More broadly, Bakker (2010: 723) defines it as where "markets determine resource allocation and pricing". While these two thinkers might present different conceptual takes on the neoliberalization of nature, across these definitions, marketization is analytically conceptualized as the 'political-economic' transformation of a natural phenomenon from an entity that wasn't subject to pricing into one that is now priced. It's notable, however, that this transformation is almost entirely 'social' rather than 'socio-natural'—that is, market pricing is conceived as alien to natural phenomena, as outside their naturalized workings.

Second, there is a tendency in the neoliberal natures literature towards valorizing nature—entailing a range of environmental processes, systems, interactions, and suchlike—as 'resisting' or 'contesting' neoliberalism, entailing the installation of markets in and across socio-natural relationships. For example, Castree (2010a: 1725) notes that neoliberalism is often "defined by its engagement with the non-human world" and especially the extent to which this "has led to a recalibration of, and even challenge to, neoliberal policies over time". Others make this point far more forcefully, noting that the limits or constraints of nature—whether as discourse or materially—represent a 'check' on unrestricted markets (e.g. McCarthy and Prudham 2004), a form of natural 'recalcitrance' (Fletcher 2014), and/or a 'fundamental challenge' (Roff 2008) to the ongoing and continuing neoliberalization of nature (see Bakker 2009). In turn, these authors argue that contestation comes to (re)configure the implementation and outcomes of neoliberal processes (e.g. privatization, marketization, corporatization, etc.). Much of this literature presents markets as an aberration of nature and of the workings of natural processes, systems, and relations in our habitats. As such, it seemingly repeats the notion that there is a distinction between social and natural phenomena, whether or not that is the intent, characterizing the latter as an important limiter of or constraint on the former, and thereby naturalizing nature as a starting condition on which humans act or have an impact.

My aim in this book is to try to push at these two assumptions—(1) that material nature is transformed by social processes and (2) that markets are disruptors of a pristine material nature. I want to analyse the market-nature hybrids that emerge as the result of the co-construction of markets and natures through neoliberalization, or otherwise. This means analysing the bio-economy—my case study here—in terms of the co-construction of the biological materialities *and* socio-political relationships that characterize its development over time. Other ways of understanding this nature-economy relationship are, therefore, worth examining at this point.

Understanding Nature-Economy Relations

For me, it is helpful to look at different ways that nature-economy relationships can be understood and conceptualized. In this section I discuss three such approaches as precursors to the general approach I take in the rest of the book, which I outline in the final section. The approaches I discuss first are (1) environmental economic geography, (2) socio-technical or sustainability transitions, and (3) political-economic materialities.

Environmental Economic Geography

Environmental economic geography (EEG) is a relatively recent sub-field in human geography that attempts to move away from treating nature as a stable, uncontested category and as "passive location, condition or resource factor input" (Heidkamp 2008: 62). According to Heidkamp (2008), economic geographers need to extend their theoretical approaches and EEG can provide the analytical tools to do so, even if it is currently a 'loose' set of concepts and positions and an 'inchoate' perspective (Bridge 2008). Although a relatively new analytical perspective, EEG has a long pedigree with origins in much earlier work on ecological modernization (Gibbs 1996). Research in EEG has so far focused on five main areas: (1) discursive and material construction of nature; (2) the diverse institutional fixes used to resolve environmental issues; (3) the proliferation of stakeholders in environmental governance; (4) whole value chain and life cycle analyses of environmental impacts; and (5) the co-constitutive interaction between natural and economic processes (e.g. Gibbs 1996, 2006; Bridge 2008; Hayter 2008; Heidkamp 2008; Hudson 2008; Soyez and Schulz 2008; Editorial 2011; Patchell and Hayter 2013; Schulz and Bailey 2014; Kedron 2015). I address each of these areas in turn.

First, EEG research highlights the discursive and material constructions of nature as a political-economic resource (e.g. Bridge 2008, 2009; Soyez and Schulz 2008). As Bridge (2009) notes, natural 'resources' (e.g. timber, woodchips, sugarcane, corn, etc.) are not found, they are made. This includes the transformation of 'waste'—a key input into the bio-economy according to many proponents—into new resources, entailing its re-valorization as a potential input into production, circulation, and consumption (Hudson 2008). While the construction of a resource is, necessarily, a material process, it's also a discursive process in that social, cultural, political, and economic visions, narratives, and discourse configure nature-economy relations (Soyez and Schulz 2008; Birch et al. 2010; Ponte and Birch 2014). For example, future visions or imaginaries of low-carbon societies are implicated in the construction of waste as a resource for the bio-economy, as several writers in other theoretical traditions illustrate (e.g. Hilgartner 2007, 2015; Levidow et al. 2012, 2013). Consequently, it's important to consider the socio-cultural dimensions of nature-economy relations as much as their materialities, since beliefs, attitudes, and preferences about or towards environmental degradation necessarily inform *how* those relations are organized and governed.

Second, EEG research emphasizes the need to understand diverse institutional fixes used to resolve environmental problems (Gibbs 2006; Soyez and Schulz 2008). While the neoliberal natures literature focuses on markets as the dominant—if not sole—institutional fix, scholars in EEG suggest that there are multiple institutional fixes at play. For example, Schulz and Bailey (2014) argue that concepts and policies like the 'green economy', 'post-growth', 'steady-state', and 'sharing economy' are also examples of institutional fixes. These types of fixes might be premised upon reducing material consumption or increasing resource efficiencies or something else entirely. However, the aim is to renovate capitalism rather than replace it, meaning that *how* the 'fix' is supposed to work is different than neoliberal, market-based instruments, even if the goal is not. As such, it is possible that major initiatives supporting societal system transformation— such as the 'green economy'—do not offer much in the way of institutional change as they first appear (Goldstein and Tyfield 2017). At least, that is, if the aim is to transform the political-economic system as well.

Third, EEG research stresses the overlapping institutional interests and conflicts that result from the proliferation of stakeholders in environmental governance. Soyez and Schulz (2008: 18) argue that there has been a "multiplication of stakeholders and relevant institutions" such as civil society groups, international organizations, governments, private sector, and so on. Not only has the number increased, but the reach of these stakeholders has also shifted as they have become increasingly international in their operations and influence. As Patchell and Hayter (2013) note, this is a consequence of many environmental problems being multi-scalar— being both 'global' and 'site specific'. Climate change immediately springs to mind here with the growing involvement of different stakeholders in things like the Conference of Parties (COP) every year. With reference to the bio-economy, however, there is a similar growth of stakeholders. For example, something like the Roundtable on Sustainable Biomaterials (RSB)—originally 'Biofuels'—represents an example of the attempts to integrate a range of stakeholders, from farmers in the Global South to petroleum refiners in the Global North. According to its website, "RSB members are organised into five chambers", reflecting the different stakeholders involved (i.e. growers, users, social, environmental, and government).[1] It is notable, though, that conflicts have also increased with the expansion of environmental analysis, standards, certification, audit, and so on leading to more contests around environmental governance (Himley 2008; Patchell and Hayter 2013).

Fourth, EEG research seeks to turn the gaze of economic geographers—and other disciplines—away from production and towards production, consumption, *and* waste. According to Gibbs (2006: 209), for example, "too much emphasis has been placed upon 'production pollution' and not enough on 'consumption pollution'" in economic geography. Gibbs makes this point in reference to the work of Ray Hudson, who also stresses the need to think about 'waste' as an important aspect of nature-economy relations (Hudson 2008). As a result, innovation and technological change are often conceived as *the* solutions to environmental problems because they solve 'production pollution', which Gibbs (2006) argues often ignores the potential impacts of consumption changes (e.g. cutting consumption). Building a theoretical framework around the 'life cycle' of goods and services—'from field to fuel' in bio-economy parlance—requires a different methodological take on nature-economy relations, including the adoption of 'value chain' methodologies (e.g. Ponte 2004, 2014; Hayter 2008). Considering the importance of energy to global political economy (e.g. electricity, transportation, internet, etc.), examining the energy use of different steps in the value chain provides one way to understand the sustainability of different nature-economy relations.

Finally, EEG research is characterized most clearly by an attempt to analyse—both theoretically and empirically—the *material* impacts and implications of environmental processes and systems on economic ones, as well as vice versa. Bridge (2008: 77) argues that political economy is not simply a question of 'value creation', it is also very much "a process of materials transformation" comprising material, energetic, and environmental transformations (also Hudson 2008). It is not simply that political economy—by which I mean the organization of production, consumption, and waste—is (re)configured as the result of material changes; the obverse is also evident—namely, nature is materially transformed as a

[1] http://rsb.org/about/who-we-are/

consequence of political economy. An Editorial (2011) in the journal *Economic Geography*, for example, stresses that this means we cannot claim that particular 'socionatures' are normatively good or bad as a result of their correspondence to some assumed notion of pristine 'nature'. Rather, understanding nature-economy relations is always necessarily a question of examining the power relations amongst actors, human and/or non-human. I come back to some of these implications below, but it's worth emphasizing that this point—that nature-economy materialities are co-constitutive—entails rethinking the way we understand innovation and technological change; these aren't processes impacting *on* nature, providing a technological fix for environmental problems, but rather processes at play in co-constructing nature *and* economy.

Innovation and Socio-technical Change

As the last point above suggests, it's important to examine innovation and socio-technical change in order to understand the relationship between nature and economy. A number of scholars working in the EEG perspective have dealt with this issue specifically and primarily from the position of wholesale changes to existing 'techno-economic paradigm' (e.g. Gibbs 2006; Hayter 2008). It's worth considering, then, *how* societies change as the result of radical systemic renewal or renovation, a topic which is the focus of an increasingly important field called 'socio-technical' or 'sustainability' transitions (e.g. Geels 2002, 2004, 2005; Geels and Schot 2007; Shove and Walker 2007; Frantzeskaki and Loorbach 2010; Geels et al. 2011; Coenen et al. 2012; Lawhon and Murphy 2012; Truffer and Coenen 2012; Tyfield 2014, 2017; Hansen and Coenen 2015). Much of this literature builds on the ideas of the historian of technology, Thomas Hughes (1983), who sought to understand how the emergence of particular scientific and technological artefacts (e.g. electric lightbulbs) is necessarily embedded within a wider set of social, political, and economic changes in societal infrastructures and institutions (e.g. electricity markets, household energy consumption, etc.).

Such socio-technical transitions represent the key way that wholesale societal change happens, according to this literature, suggesting that any understanding of the bio-economy—itself premised on significant social and technical change—necessarily entails an analysis of the transition process. An apt example to illustrate this point is the rise of the automobile in early- to mid-twentieth-century 'Western' culture (Geels 2005;

Geels et al. 2011) and its subsequent dissemination globally to emerging economic powers like China (Tyfield 2017). According to Geels (2005), automobiles replaced horse-drawn carriages as the result of the alignment of elements in the socio-technical system, including the artefacts themselves (e.g. car) as well as the knowledges (e.g. driving), users (e.g. drivers), markets (e.g. car manufacturers), regulations (e.g. road safety), culture (e.g. individualism), infrastructures (e.g. highways), and so on, around them. As one socio-technical system emerges, and another disappears, the social *and* material landscape itself is reconfigured through changes to settlements (e.g. suburbs), travel (e.g. highways), mobility (e.g. commuting), energy production (e.g. oil), and culture (e.g. road movies) (see Mitchell 2011; Huber 2013). Today, automobile markets are now thoroughly enmeshed within fossil fuel energy regimes, which can be conceived as simultaneously social *and* material systems. How alternative energies like biofuels might challenge, integrate, or diversify this socio-technical system is precisely why I'm writing this book.

At present, this sustainability transitions literature is dominated by the work of people like Frank Geels (2002) and his collaborators. Using the 'multi-level perspective' (or MLP), Geels and others are concerned with understanding how one socio-technical 'regime' is replaced by another in systems transitions. Within the MLP, a regime is comprised of "a cluster of elements, including technology, regulations, user practices and markets, cultural meanings, infrastructure, maintenance networks, and supply networks" (Geels 2005: 681). It's part of a wider multi-scalar and multi-actor process in which certain regimes (e.g. automobiles) are actively stabilized in response to emerging and competing threats from various 'niche' innovations (e.g. electric cars, biofuels, etc.). All of this sits within a broader 'landscape' of social, cultural, and political pressures or forces that are not part of particular regimes or niches; for example, global climate change policy. Both niche innovations and landscape pressures can lead to regime change according to Geels (2002). It is notable that this theoretical approach is very popular, although it is perhaps also notable that the sustainability transitions field is predominantly embedded within European scholarly debates and is largely absent from North America—with exceptions (e.g. Birch 2016b; Gliedt et al. 2018).

Although there is a focus on *socio-technical* systems, transitions, innovation, and change, it is perhaps as useful to focus more explicitly on the *socio-material* aspects of these things—as I've argued elsewhere (e.g. Birch 2016b). This is for several reasons. First, much of the conceptual discussion about regimes (and niches) focuses on socio-economic aspects of systems—that is, the associated "technology, regulations, user practices and markets, cultural meanings, infrastructure, maintenance networks, and supply networks" (Geels 2005: 681)—rather than the physical materiality. In fact, the landscape—in both material and social terms—is largely treated as an 'out-there' force, rather than a directly implicated and constitutive aspect of nature-economy relations. Second, socio-technical systems are necessarily geographical, as a number of human geographers have stressed (e.g. Coenen et al. 2012; Truffer and Coenen 2012; Hansen and Coenen 2015; Calvert et al. 2017a). They have sought to integrate geographical processes into the sustainability transitions oeuvre. For example, Truffer and Coenen (2012: 10) argue that the sustainability transitions literature tends to conceptualize institutions (e.g. markets, regulation) in abstract terms, whereas it's important to consider the "interdependencies among institutional configurations in specific places". Hansen and Coenen (2015) make a similar argument, noting that it is important to consider the geographical specificity of inter-organizational relations to transitions. One particularly critical issue to consider with the bio-economy, for example, is *how* prevailing and dominant regimes (e.g. fossil fuels) are disrupted across geographically specific infrastructures (e.g. distribution routes), institutions (e.g. energy markets), value chains (e.g. feedstock sourcing), and social values (e.g. driving) (Calvert et al. 2017b). Finally, socio-technical transitions are characterized as much by material transformations as they are by social and technical ones—the issue I turn to now.

Political-Economic Materialities

As I sought to illustrate in the discussion of the neoliberal natures literatures above, there is a tendency—intentional or otherwise—to position the economy as an imposition on nature, assuming that the former, in its neoliberal guise especially, is corrupting nature, or is an aberration of nature. I think this analytical position is not tenable and that it's important to get to grips with the socio-material dimensions of the nature-economy relation in order to avoid this tendency myself. A good starting point is the recent and ongoing work in science and technology studies (STS)—broadly conceived—which argues that economies and markets are 'material' assemblages or entities (e.g. Callon 1998; Pinch and Swedberg 2008; MacKenzie 2009; Mitchell 2009, 2011). Echoing the arguments made in environmental economic geography (EEG), this material markets literature takes 'materiality' seriously in considering the interplay between social and biophysical worlds. My focus here is on the work of Timothy Mitchell who provides a useful starting point for thinking about the *political-economic materialities* of bioenergy (see Birch and Calvert 2015).

Mitchell is primarily concerned with how the materialities of different forms of fossil energy enable or constrain different forms of political mobilization and movements, which I'm going to broaden to political-economic processes. Mitchell's basic point is that the biophysical characteristics of coal in the nineteenth century led to the rise of mass democratic movements in countries like Britain, largely based on labour organizing and as a direct result of the material conditions of coal extraction and distribution. According to Mitchell (2011), coal was dependent on a large labour force which controlled several choke points in the energy system (e.g. coal pits, harbours, train yards, etc.) and meant that workers had the power to shut down energy supply. As a result of this political materiality, broad-based political movements emerged through labour organizing, universal suffrage, and so on. In contrast, however, the materiality of the twentieth-century oil energy system limited the capacity of workers to take collective action because oil is drilled and piped, meaning that it requires fewer workers to extract and transport. Consequently, this explains the so-called resource curse that often afflicts oil-rich countries, especially in the Global South, where the returns on oil wealth are not shared equitably or socially.

Kirby Calvert and I have applied Mitchell's theory to bioenergy (Birch and Calvert 2015). We were interested in whether the biophysical materialities of bioenergy help explain the development of the biofuels sector in Canada (see Chap. 5). Our argument extends the notion of 'political' materialities to 'political-economic' materialities by considering how the biophysical characteristics of bioenergy enable and constrain the deployment of biofuels as an alternative transport fuel. We argue that energy density, feedstock immobility, and trans-boundary emissions calculations all help to explain the focus on so-called 'drop-in' biofuels, which are supposed to slot into existing energy infrastructure and institutions (e.g. pipelines, refineries, markets, etc.). Biofuels aren't, in the Canadian context at least, meant to destabilize or challenge incumbent industries like the oil and gas sector. Innovation, such as it is, is framed by this focus on dropping in, reinforcing the prevailing political-economic materialities of the fossil fuel system—rather than engendering any change. In light of these findings, it's important to remember two things. First, innovation is not only a social and technical transformation, it's also a material transformation that (re)configures infrastructures, institutions, value chains, and physical environments as part of a wider systemic change (Birch 2013, 2016a, b; Becker et al. 2016). Second, there are political-economic materialities to innovation or technological change, where certain materialities enable and constrain certain research, development, and commercialization processes.

THE CO-CONSTRUCTION OF MARKETS AND NATURES

Analytically, my aim in this book is to bring this range of different perspectives focused on nature-economy relations to bear on the emergence of the bio-economy as a social phenomenon. My starting point is the notion that the bio-economy represents the neoliberalization of nature and our associated understanding of nature, which I've argued in the past (e.g. Birch 2006; Birch et al. 2010). That is the reason I started this theoretical discussion with the neoliberal natures literature, which is based on the notion that the market has been installed as the key societal institution for organizing nature-economy relations (and much else besides). However, I think that this theoretical perspective ends up limiting our understanding of the bio-economy, primarily because it presents nature-economy relations in a particular way. As noted above, the concept of neoliberalism, including as a process (i.e. neoliberalization), is underpinned by an inherent critique of markets (Storper 2016), treating them as almost an aberration of nature.

Without wishing to offer support to a 'neoliberal' position, it's important to analyse how markets are embedded within and bounded by a range of social, technical, *and* material relations, which precludes treating them like some sort of distortion of those same relations.[2] All of this is framed, theoretically, by a Polanyian understanding of markets as 'instituted process' (Polanyi 1944), meaning that when we talk about markets, we're necessarily talking about the state as well, since the state is active in enacting those markets. Needless to say, markets come to underpin state rationales as well, but I don't want to get caught in some sort of recursive loop, unpacking co-construction all the way down.

---

[2] There is an interesting similarity here with neoclassical economics which treats markets as analytically natural and, therefore, any interference in their operation as a distortion.

Focusing on nature-economy relations in this conceptual framing does not mean accepting that markets are *the* solution to environmental problems; rather it means examining how the co-construction of markets and natures comes to produce—and/or, it's important to note, reduce—those problems. As such, I'm trying to avoid the analytical assumption that natures and markets are or can be separate/d from one another, even normatively speaking, and that, therefore, we cannot naturalize either as a pristine starting point which ends up corrupted by the other. Similarly, I'm trying to avoid the analytical assumption that natures and markets are inherently in conflict, especially the notion that non-human materialities challenge or contest the imposition of markets upon their biophysical form. Rather, my aim in this book is to analyse how natures and markets might be considered as co-constructed, and to do this necessitates understanding different aspects of the (re)configuration of nature-economy relations.

First, it means analysing the different discourses we use to frame our actions in the world (Soyez and Schulz 2008; Ponte and Birch 2014). How do we narrate our lives, our actions, and our decisions to make sense of them and to make them sensible? This necessitates examining the role of future narratives, visions, and imaginaries that we produce, and trying to understand what effects they might have. Several scholars have already highlighted the importance of these imaginaries in promoting and supporting the emergence of the bio-economy (e.g. Hilgartner 2007, 2015; Birch et al. 2010; Levidow et al. 2012; Birch 2016a, b; Bauer 2018). Empirically, it means trying to understand how these imaginaries (re)configure the standards, supply chains, infrastructures, institutions, and consumer attitudes that underpin the development of advanced biofuels—and vice versa.

Second, it means analysing the different political-economic stakeholders implicated in the emergence and governance of the bio-economy (Himley 2008; Patchell and Hayter 2013). How do we understand government and industry attempts to create markets in the bio-economy? Markets do not just exist, they are instituted or created through deliberate and, sometimes, deliberative action. As such, it's necessary to examine the specific 'market development' policies that government, industry, and civil society pursue or support. For example, conventional and advanced Biofuels production is dependent upon national or sub-national mandates and targets, whether those are volumetric (e.g. the USA) or proportional (e.g. the EU). More generally, the bio-economy is dependent on a range of enacted policies, including mandates and targets mentioned already, standards and criteria to establish sustainability benefits, labelling and certification to increase visibility of bio-based products and services, and feedstock supply contracts to establish political-economic certainty needed to attract investment. This is how markets are made.

Third, it means analysing the biophysical materialities of the bio-economy (Mitchell 2009, 2011; Birch and Calvert 2015; Becker et al. 2016). How do we understand the implications of these materialities to the legitimation of particular political-economic decisions? 'Nature' is often used as a shorthand for 'should', a normative shortcut that we can use in our arguments or decisions. Materiality has a legitimating authority to it, one that constrains and enables certain actions as Mitchell (2011) notes. For example, one of the most attractive aspects to conventional biofuels is that they do not a require significant reconfiguration of the existing energy regime. Moreover, advanced biofuels are premised on the idea that they can 'drop-in' to existing infrastructures and institutions. While this might make biofuels an attractive pathway to pursue towards a low-carbon future, there are downsides too. In particular, conventional and advanced biofuels are, by and large, thoroughly entangled with existing fossil energy regimes, to the extent that they can be described as 'correlated'—that is, the economic viability of biofuels is tied to fossil energy prices.

Fourth, it means bringing the political-economic *and* material aspects together to understand what sort of biotechnological innovation has emerged as the result of the co-construction of markets and natures. How do we understand the interplay between and outcomes from the co-construction of markets and natures? Understanding the co-construction of markets and natures involves examining the effects of the interaction between political-economic and environmental processes. For example, when it comes to identifying suitable feedstock for advanced biofuels, there is an emphasis on finding ways to convert heterogenous inputs (e.g. organic waste, timber offcuts, energy crops, agricultural castoffs, etc.), which have very different biophysical characteristics, into a valuable and homogenous output (e.g. biofuel). However, this innovation process is dependent on constructing these inputs as material 'residues' which have no market 'value'. As I discuss later, the problem with this co-construction is that residues can easily *acquire* value as their use and demand increases.

Finally, it means examining the potential of the bio-economy as a socio-material pathway towards a low-carbon future. How might the bio-economy contribute to societal transitions from this perspective? As much as we might want the bio-economy to be a potential transition pathway, it's not always clear what it represents. For example, and in contrast to much of the sustainability transitions literature (e.g. Geels 2002, 2004; Geels and Schot 2007; Geels et al. 2011), the bio-economy cannot be considered as an emerging 'regime' that will replace an existing regime. Instead, and as I'll show throughout the rest of the book, where the bio-economy is presented as anything more than a simple substitution of bio-based materials for existing fossil fuel ones, it's premised on the wholesale transformation of social, political, and economic infrastructures and institutions. Something supposedly simply like feedstock supply entails a number of dramatic socio-material changes, including large-scale and constant transportation of biomass from harvest sites to processing refineries; mass distribution of processing refineries into communities; and a complex accounting of who can and who cannot claim credit for carbon emissions reductions (Birch and Calvert 2015; Calvert et al. 2017b). It is, then, important to think about how the dominant bio-economy imaginary—present in most policy visions—could be transformed into an alternative vision based on broader concerns with 'circular' or 'life-cycle' political-economic provisioning.

CONCLUSION

In this chapter, I've sought to lay out my theoretical approach for understanding 'neoliberal bio-economies'. It was necessary to start with an outline of what I meant by neoliberalism and the theoretical approaches used in the analysis and critique of it. In particular, I draw on the processual approach common to debates in human geography (e.g. Peck and Tickell 2002), especially as this relates to the analysis of the neoliberalization of nature. There is a considerable literature on these so-called neoliberal natures, so I drew on a series of reviews by people like McCarthy and Prudham (2004), Castree (2008a, b, 2010a, b), and Bakker (2009, 2010). My aim was to highlight the conceptual starting point for many analyses of nature-economic relations. As I sought to show, these analyses tend to present markets as inherently problematic and as disrupting or distorting environmental systems, processes, and habitats. Rather than take this analytical starting point, however, I wanted to consider other perspectives.

In thinking through nature-economy relations, I drew on three analytical traditions. First, I discussed the emerging field of environmental economic geography, or EEG. According to many EEG researchers, there is a need to think through the discursive and material constructions of nature and natural resources; the expansion of institutional fixes for environmental problems; the proliferation of stakeholders and stakeholder conflicts that results from this expansion; the need to consider the full life cycle of production, consumption, and waste; and the co-construction of economic and environmental processes (e.g. Heidkamp 2008; Soyez and Schulz 2008; Bridge 2008; Patchell and Hayter 2013). Second, I discussed the burgeoning field of socio-technical or sustainability transitions. Although it offers valuable insights into societal transformations and the role of innovation in that process (e.g. Geels 2002, 2004), the focus of this literature on social and technical aspects of this process obscures the impacts of the material characteristics of non-human entities on transitions, which brings me to the final tradition I discussed, namely, that of political-economic materialities (e.g. Mitchell 2009, 2011). Introducing materialities into the mix helps to understand how certain innovation processes or *socio-material* pathways are enabled and others constrained.

REFERENCES

Bailey, I. (2007a) Market environmentalism, new environmental policy instruments, and climate policy in the United Kingdom and Germany, *Annals of the Association of American Geographers* 97(3): 530–550.

Bailey, I. (2007b) Neoliberalism, climate governance and the scalar politics of EU emissions trading, *Area* 39(4): 431–442.

Bakker, K. (2009) Commentary: Neoliberal nature, ecological fixes, and the pitfalls of comparative research, *Environment and Planning A* 41: 1781–1787.

Bakker, K. (2010) The limits of 'neoliberal natures': Debating green neoliberalism, *Progress in Human Geography* 34(6): 715–735.

Bauer, F. (2018). Narratives of biorefinery innovation for the bioeconomy – conflict, consensus, or confusion? *Environmental Innovation and Societal Transitions*, doi.org/10.1016/j.eist.2018.01.005.

Bekker, S., Moss, T. and Naumann, M. (2016) The importance of space: Towards a socio-material and political geography of energy transitions, in L. Gailing and T. Moss (eds) *Conceptualizing Germany's energy transition*, London: Palgrave Macmillan, pp. 93–108.

Birch, K. (2006) The neoliberal underpinnings of the bioeconomy: The ideological discourses and practices of economic competitiveness, *Genomics, Society and Policy* 2(3): 1–15.

Birch, K. (2013) The political economy of technoscience: An emerging research agenda. *Spontaneous Generations: A Journal for the History and Philosophy of Science* 7 (1): 49–61

Birch, K. (2015) *We have never been neoliberal: A manifesto for a doomed youth*, Winchester: Zero Books.

Birch, K. (2016a) Emergent policy imaginaries and fragmented policy frameworks in the Canadian bio-economy, *Sustainability* 8(10): 1–16.

Birch, K. (2016b) Materiality and sustainability transitions: Integrating climate change into transport infrastructure in Ontario, Canada, *Prometheus: Critical Studies in Innovation* 34(3–4): 191–206.

Birch, K. (2017) *A research agenda for neoliberalism*, Cheltenham: Edward Elgar.

Birch, K. and Calvert, K. (2015) Rethinking 'drop-in' biofuels: On the political materialities of bioenergy, *Science and Technology Studies* 28: 52–72.

Birch, K. and Mykhnenko, V. (eds) (2010) *The rise and fall of neoliberalism*, London: Zed Books.

Birch, K. and Siemiatycki, M. (2016) Neoliberalism and the geographies of marketization: The entangling of state and markets, *Progress in Human Geography* 40(2): 177–198.

Birch, K., Levidow, L. and Papaioannou, T. (2010) *Sustainable Capital?* The neoliberalization of nature and knowledge in the European knowledge-based bioeconomy, *Sustainability* 2(9): 2898–2918.

Birner, R. (2018) Bioeconomy concepts, in I. Lewandowski (eds) *Bioeconomy*, Cham: Springer, pp. 17–38.

Bonneuil, C. and Fressoz, J-B. (2017) *The shock of the anthropocene*, London: Verso.

Bowman, S. (2016) I'm a neoliberal. Maybe you are too, *medium.com*, retrieved from: https://medium.com/@s8mb/im-a-neoliberal-maybe-you-are-too-b809a2a588d6#.ffj5mgbbo.

Brenner, N., Peck, J. and Theodore, N. (2010) Variegated neoliberalization: Geographies, modalities, pathways, *Global Networks* 10(2): 182–222.

Bridge, G. (2008) Environmental economic geography: A sympathetic critique, *Geoforum*, 39, 76–81.

Bridge, G. (2009) Material worlds: Natural resources, resource geography and the material economy, *Geography Compass* 3(3), 1217–1244.

Brown, W. (2015) *Undoing the demos*, Cambridge, MA: Zone Books.

Büscher, B., Dressler, W. and Fletcher, R. (eds) (2014). *Nature™ Inc: Environmental conservation in the neoliberal age*, Tucson, AZ: University of Arizona Press.

Cahill, D., Cooper, M., Konings, M. and Primrose, D. (eds) (2018) *The SAGE handbook of neoliberalism*, London: SAGE.

Callon, M. (ed.) (1998) *The laws of the markets*, Oxford: Blackwell

Calvert, K., Kedron, P., Baka, J. and Birch, K. (2017a) Geographical perspectives on sociotechnical transitions and emerging bio-economies: Introduction to a special issue, *Technology Analysis & Strategic Management* 29(5): 477–485.

Calvert, K., Birch, K. and Mabee, W. (2017b) New perspectives on an old energy resource: Biomass and emerging bio-economies, in S. Bouzarovski,

M. Pasqualetti and V. Castan Broto (eds) *The Routledge research companion to energy geographies*, London: Routledge, pp. 47–60.

Castree, N. (2006) From neoliberalism to neoliberalisation: Consolations, confusions, and necessary illusions, *Environment and Planning A* 38: 1–6.

Castree, N. (2008a). Neoliberalising nature: The logics of deregulation and reregulation, *Environment and Planning A* 40: 131–152.

Castree, N. (2008b) Neoliberalising nature: Processes, effects, and evaluations, *Environment and Planning A* 40: 153–173.

Castree, N. (2010a) Neoliberalism and the biophysical environment 1: What 'neoliberalism' is, and what difference nature makes to it, *Geography Compass* 4(12): 1725–1733.

Castree, N. (2010b) Neoliberalism and the biophysical environment 2: Theorising the neoliberalisation of nature, *Geography Compass* 4(12): 1734–1746.

Club of Rome. (1972) *The limits to growth*, New York: Universe Books.

Coenen, L., Benneworth, P. and Truffer, B. (2012) Towards a spatial perspective on sustainability transitions, *Research Policy* 41(6): 968–979.

Collard, R-C., Dempsey, J. and Rowe, J. (2016) Re-regulating socioecologies under neoliberalism, in S. Springer, K. Birch and J. MacLeavy (eds) *The handbook of neoliberalism*, New York: Routledge, pp. 455–465.

Dardot, P. and Laval, C. (2014) *The new way of the world*, London: Verso.

Dempsey, J. and Robertson, M. (2012) Ecosystem services: Tensions, impurities, and points of engagement within neoliberalism, *Progress in Human Geography* 36(6): 758–779.

Duménil, G. and Lévy, D. (2004) *Capital resurgent*, Cambridge MA: Harvard University Press.

Dunn, B. (2017) Against neoliberalism as a concept, *Capital and Class* 41(3): 435–454.

Editorial. (2011) Emerging themes in economic geography: Outcomes of the Economic Geography 2010 Workshop, *Economic Geography* 87(2): 111–126.

Essex, J. (2008) Biotechnology, sound science, and the Foreign Agricultural Service: A case study in neoliberal rollout, *Environment and Planning C* 26: 191–209.

Essex, J. (2016) The neoliberalization of agriculture: Regimes, resistance, and resilience, in S. Springer, K. Birch, and J. MacLeavy (eds.) *The handbook of neoliberalism*, New York: Routledge, pp. 514–525.

Fletcher, R. (2014) Taking the chocolate laxative: Why neoliberal conservation "fails forward", in B. Büscher, W. Dressler, and R. Fletcher (eds) *Nature™ Inc: Environmental conservation in the neoliberal age*, Tucson, AZ: University of Arizona Press, pp. 87–107.

Foucault, M. (2008) *The birth of biopolitics*, New York: Picador.

Frantzeskaki, N. and Loorbach, D. (2010) Towards governing infrasystem transitions, *Technological Forecasting and Social Change* 77(8): 1292–1301.

Gareau, B. (2008) Dangerous holes in global environmental governance: The roles of neoliberal discourse, science, and California agriculture in the Montreal Protocol, *Antipode* 40(1): 102–130.

Geels, F. (2002) Technological transitions as evolutionary reconfiguration processes: A multi-level perspective and case study, *Research Policy* 31: 1257–1274.

Geels, F. (2004) Understanding system innovations: A critical literature review and a conceptual synthesis', in B. Elzen, F. Geels and K. Green (eds.), *System innovation and the transition to sustainability: Theory, evidence and policy*, Cheltenham: Edward Elgar, pp. 19–47.

Geels, F. (2005) The dynamics of transitions in socio-technical systems: A multi-level analysis of the transition pathway from horse-drawn carriages to automobiles (1860–1930), *Technology Analysis & Strategic Management* 17(4): 445–476.

Geels, F., Kemp, R., Dudley, G. and Lyons, G. (2011) *Automobility in transition?: A socio-technical analysis of sustainable transport*, London: Routledge.

Geels, F. and Schot, J. (2007) Typology of sociotechnical transition pathways, *Research Policy* 36(3): 399–417.

Gibbs, D. (1996) Integrating sustainable development and economic restructuring: A role for regulation theory? *Geoforum*, 27(1), 1–10.

Gibbs, D. (2006) Prospects for an environmental economic geography: Linking ecological modernization and regulationist approaches. *Economic Geography*, 82(2), 193–215.

Gliedt, T., Hoicka, C. and Jackson, N. (2018) Innovation intermediaries accelerating environmental sustainability transitions, *Journal of Cleaner Production* 174: 1247–1261.

Goldstein, J. and Tyfield, D. (2017). Green Keynesianism: Bringing the entrepreneurial state back in(to question)? *Science as Culture*. doi.org/10.1080/09505431.2017.1346598.

Hansen, T. and Coenen, L. (2015) The geography of sustainability transitions: Review, synthesis and reflections on an emergent research field, *Environmental Innovation and Societal Transitions* 17: 92–109.

Harvey, D. (2005) *A brief history of neoliberalism*, Oxford: Oxford University Press.

Hayter, R. (2008) Environmental economic geography. *Geography Compass*, 2, 1–20.

Heidkamp, P. (2008) A theoretical framework for a 'spatially conscious' economic analysis of environmental issues, *Geoforum* 39: 62–75.

Hilgartner, S. (2007) Making the bioeconomy measurable: Politics of an emerging anticipatory machinery, *BioSocieties* 2 (3): 382–386.

Hilgartner, S. (2015) Capturing the imaginary: Vanguards, visions, and the synthetic biology revolution, in S. Hilgartner, C. Miller and R. Hagendijk (eds) *Science and democracy: Making knowledge and making power in the biosciences and beyond*, London: Routledge, pp. 33–55.

Himley, M. (2008) Geographies of environmental governance: The nexus of nature and neoliberalism, *Geography Compass* 2(2): 433–451.

Huber, M. (2013) *Lifeblood*, Minneapolis: University of Minnesota Press.

Hudson, R. (2008) Cultural political economy meets global production networks: A productive meeting?, *Journal of Economic Geography* 8(3): 421–440.

Hughes, T. (1983) *Networks of power*, Baltimore: John Hopkins University Press.

Jessop, B. (2016) The heartland of neoliberalism and the rise of the austerity state, in S. Springer, K. Birch and J. McLeavy (eds) *The handbook of neoliberalism*, New York: Routledge, pp. 410–421.

Kedron, P. (2015) Environmental governance and shifts in Canadian biofuel production and innovation, *The Professional Geography* 67(3): 385–395.

Kenney-Lazar, M. and Kay, K. (2017) Value in capitalist natures, *Capitalism Nature Socialism* 28(1): 33–38.

Larner, W. (2003) Neoliberalism? *Environment and Planning D* 21: 509–12.

Lawhon, M. and Murphy, J. (2012) Socio-technical regimes and sustainability transitions: Insights from political ecology. *Progress in Human Geography* 36(3): 354–378.

Levidow, L., Birch, K. and Papaioannou, T. (2012) EU agri-innovation policy: Two contending visions of the bio-economy, *Critical Policy Studies* 6 (1): 40–65.

Levidow, L., Birch, K. and Papaioannou, T. (2013) Divergent paradigms of European agro-food innovation: The knowledge-based bio-economy (KBBE) as an R&D agenda, Science, *Technology and Human Values* 38(1): 94–125.

Loftus, A. and Budds, J. (2016) Neoliberalizing water, in S. Springer, K. Birch and J. McLeavy (eds) *The handbook of neoliberalism*, New York: Routledge, pp. 503–513.

Lohmann, L. (2010) Neoliberalism and the calculable world: The rise of carbon trading, in K. Birch and V. Mykhnenko (eds) *The rise and fall of neoliberalism*, London: Zed Books, pp. 77–93.

MacKenzie, D. (2009) *Material markets*, Oxford: Oxford University Press.

McAfee, K. (2003) Neoliberalism on the molecular scale: Economy and genetic reductionism in biotechnology battles, *Geoforum* 34: 203–219.

McCarthy, J. and Prudham, S. (2004) Neoliberal nature and the nature of neoliberalism, *Geoforum* 35: 275–283.

Mirowski, P. (2013) *Never let a serious crisis go to waste*, Cambridge, MA: Harvard University Press.

Mitchell, T. (2009) Carbon democracy, *Economy and Society*, 38(3): 399–432.

Mitchell, T. (2011) *Carbon democracy*, London: Verso.

Moore, J.W. (2015) *Capitalism in the web of life*, London: Verso.

Patchell, J. and Hayter, R. (2013) Environmental and evolutionary economic geography: Time for EEG[2]? *Geografiska Annaler B* 95(2): 111–130.

Peck, J. (2010) *Constructions of neoliberal reason*, Oxford: Oxford University Press.

Peck, J. and Tickell, A. (2002) Neoliberalizing space, *Antipode* 34(3): 380–404.

Pinch, T. and Swedberg, R. (eds) (2008) *Living in a material world*, Cambridge MA: MIT Press.

Polanyi, K. (1944 [2001]). *The great transformation*, New York: Beacon Press.

Ponte, S. (2004) *Standards and sustainability in the coffee sector*, Manitoba: International Institute for Sustainable Development.

Ponte, S. (2014) The evolutionary dynamics of biofuel value chains: From unipolar and government-driven to multipolar governance, *Environment and Planning A* 46(2): 353–372.

Ponte, S. and Birch, K. (2014) Introduction: The imaginaries and governance of 'biofueled futures', *Environment and Planning A* 46(2): 271–279.

Prudham, S. (2005) *Knock on wood: Nature as commodity in Douglas-Fir country*, New York: Routledge.

Prudham, S. (2007) The fictions of autonomous invention: Accumulation by dispossession and life patents in Canada, *Antipode* 39(3): 406–429.

Roff, R.J. (2008) Preempting to nothing: Neoliberalism and the fight to de/re-regulate agricultural biotechnology, *Geoforum* 39: 1423–1438.

Schulz, C. and Bailey, I. (2014) The green economy and post-growth regimes: Opportunities and challenges for economic geography, *Geografiska Annaler B* 96(3): 277–291.

Shove, E. and Walker, G. (2007) CAUTION! Transition ahead: Politics, practice, and sustainable transition management, *Environment and Planning A* 39(4): 763–770.

Sidaway, J. and Hendrikse, R. (2016) Neoliberalism version 3.0, in S. Springer, K. Birch and J. McLeavy (eds), *The handbook of neoliberalism*, New York: Routledge, pp. 574–582.

Soyez, D. and Schulz, C. (2008) Editorial: Facets of an emerging environmental economic geography (EEG), *Geoforum*, 39, 17–19.

Springer, S., Birch, K. and MacLeavy, J. (eds) (2016) *The handbook of neoliberalism*, New York: Routledge.

Storper, M. (2016) The neo-liberal city as idea and reality, *Territory, Politics, Governance* 4(2): 241–263.

Tickell, A. and Peck, J. (2003) Making global rules: Globalisation or neoliberalisation? in J. Peck and H. Yeung (eds) *Remaking the global economy*, London: SAGE, pp. 163–182.

Truffer, B. and Coenen, L. (2012) Environmental innovation and sustainability transitions in regional studies, *Regional Studies* 46(1): 1–21.

Tyfield, D. (2014) Putting the power in 'socio-technical regimes' – E-mobility transition in China as political process, *Mobilities* 9(4): 585–603.

Tyfield, D. (2017) *Liberalism 2.0 and the rise of China*, London: Routledge.

Venugopal, R. (2015) Neoliberalism as concept, *Economy and Society* 44(2): 165–187.

Wainwright, J. and Mann, G. (2013) Climate leviathan, *Antipode* 45(1): 1–22.

World Commission on Environment and Development. (1987) *Our common future*, Oxford: Oxford University Press.

CHAPTER 3

Background to Emerging Bio-Economies

Introduction

The transition to a low-carbon future is essential if we want to avoid the worst effects of climate change. What those effects might be is hard to predict, entailing a necessary adjustment to uncertainty and indeterminacy in our decision-making. According to Diana Liverman (2009), the international response to climate change is driven by the idea that climate change is 'dangerous', that different countries have different levels of responsibility for climate change, and that markets are the best instrument for dealing with climate change. For example, the 1992 United Nations Framework Convention on Climate Change (UNFCCC) and its extension through the 1997 Kyoto Protocol reflect this underlying rationale. Signatories to the Kyoto Protocol committed to reducing their greenhouse gas (GHG) emissions below 1990 levels; this was five percent below in the 2008–2012 commitment period and 18 percent below in the 2013–2020 commitment period.[1] As a primarily market-based approach, Kyoto has been reinforced subsequently at the 2007 Bali Conference of Parties (COP13) and again at the 2009 Copenhagen (COP15) and 2015 Paris (COP21) UNFCCC meetings. As a result, the international response to climate change has been defined as a 'neoliberal' approach to dealing with climate change (e.g. Bailey 2007; Lohmann 2010, 2016).

---

[1] http://unfccc.int/kyoto_protocol/items/2830.php

© The Author(s) 2019   45
K. Birch, *Neoliberal Bio-Economies?*,
https://doi.org/10.1007/978-3-319-91424-4_3

Although various GHG emissions are released into the Earth's atmosphere, concerns with anthropogenic climate change focus on emissions of carbon dioxide (CO_2) and the build-up and concentration of CO_2 in the atmosphere; these are referred to as gigatons released per year and parts per million (ppm) in the atmosphere respectively.[2] Current CO_2 emissions surpassed 30 gigatons per year in 2011 and carbon concentration surpassed 400 ppm in 2016.[3] According to Bill McKibben (2012), we can only release about 565 more gigatons by 2050 if we want to stay below a 2 °C global temperature rise. However, at COP21 in 2015, the participating countries agreed to aim for a 1.5 °C rise, meaning that we can now only release another 353 gigatons of carbon (McKibben 2016). And this is where we come up against a major problem; namely, we have far more carbon in fossil fuel reserves around the world than 353 gigatons—close to three times as much. Carbon lock-in is a very real concern in light of these figures (Unruh 2000; Calvert 2016).

We have to make choices at this point *if* we want to avoid global temperature rises, and some people seem not to care (Oreskes and Conway 2010). People like McKibben have started campaigns like 350.org to disinvest from fossil fuel companies, while others propose grander schemes centred on a 'Green New Deal' (Jackson 2009; Goldstein and Tyfield 2017; Goldstein 2018). All these different campaigns, schemes, policies, and suchlike have a common focus on transforming our societies and economies into low-carbon versions of themselves. And this is where something like the bio-economy enters the picture. It's presented as a potential transition pathway to a low-carbon future, based on replacing fossil fuels with biological material as the key input into energy, manufacturing, chemicals, agriculture, and so on. Starting with the search for alternative energy sources (e.g. liquid biofuels) to fossil fuels, especially oil use in transport, the bio-economy is increasingly presented as a viable option for countries to take in transforming their economies. Its history can be traced back to the mid-2000s when the Organisation for Economic Development and Co-operation (OECD) and European Commission (EC) developed policy agendas supporting the idea.

This chapter provides readers with an introduction to the emergence of this bio-economy concept and policy agenda; it's going to be largely descriptive, leaving the analysis for later chapters. I first discuss climate change and the attempts to find solutions to rising fossil energy use that is

[2] http://www.ipcc.ch/publications_and_data/ar4/syr/en/contents.html
[3] https://www.esrl.noaa.gov/gmd/ccgg/trends/global.html

the primary cause of carbon emissions. Biofuels represent a key solution pursued by various governments around the world, especially advanced biofuels produced from non-food crops and materials (e.g. organic waste, residues, switchgrass, algae, etc.). These advanced biofuels represent the main object of analysis in the remaining chapters. After this, I discuss definitions of the bio-economy as a concept and the development and current state of bio-economy policy visions and strategies around the world, outlining some of the different approaches taken by different countries.

CLIMATE CHANGE AND LOW-CARBON FUTURES

Carbon Emissions and Changing Climate

The nineteenth-century scientist John Tyndall was the first person to demonstrate that the Earth's atmosphere acted like a large greenhouse. The so-called greenhouse effect was the result of atmospheric gases trapping heat, thereby warming the planet's surface. A major contributor to this greenhouse effect is carbon dioxide, or CO_2. As Robbin et al. (2010: 139) note, carbon "is constantly circulated into life forms and also captured in oceans and the atmosphere". Carbon underpins the Earth's living systems, including plant and animal life, as well as the geological formations (e.g. sedimentary rocks) and deposits (e.g. oil) that result from them.

As humans have extracted oil, coal, and other fossilized materials as the result of industrialization, we've released this carbon into the atmosphere "at a rate far faster than they are naturally returned to geological storage" (ibid.; emphasis in original). For example, according to the Stern Review (HM Treasury 2006), the stock of greenhouse gases (GHGs) in the atmosphere rose from 280 ppm CO_2 before the Industrial Revolution to 430 ppm in the mid-2000s. See Fig. 3.1 for details of global carbon emissions since 1959, distinguishing between different macro-regions around the world. As this graph illustrates, there has been a significant rise in GHGs since the 1950s, with particular growth in Asia from the 1980s onwards; in general, the graph shows that GHGs have been growing in the Global North and falling in the Global South. More specific to this book, Canadian GHG emissions reached a high of 749 Mt. CO_2 equivalent in 2007 before falling to 699 in 2012 (Wood and Mabee 2016).

As a result of increasing carbon emissions from human activity, the planet's average temperature is rising. According to NASA, average global surface temperature has risen by around 0.8 °C since 1880 with two-thirds

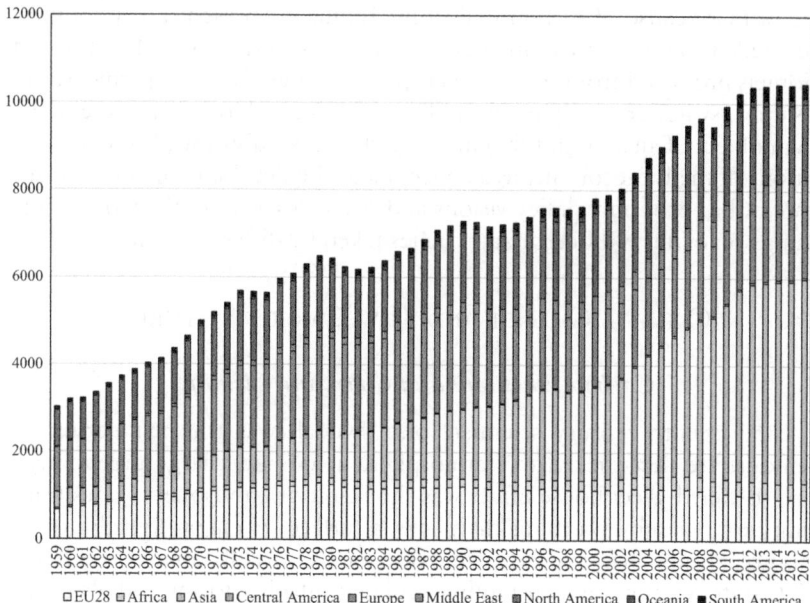

Fig. 3.1 Fossil fuel and cement production emissions. (Note: Figure is million tonnes of carbon per year (MtC/yr), where 1MtC = 1 million tonne of carbon = 3.664 million tonnes of CO_2. Source: UNFCCC (June 2017) and CDIAC, Le Quéré et al. (2017))

of that happening since the mid-1970s.[4] People use the phrase 'anthropogenic global warming' to refer to this phenomenon because it involves rising global temperature caused specifically by human activities (e.g. burning fossil fuels in large quantities). An increasing number of people are also now using the term 'anthropocene' to refer to humanity's geological impact (Bonneuil and Fressoz 2017), or 'capitalocene' to refer to the specific impacts of capitalism on the environment (Moore 2015).

The big fear is that the average global temperature will continue to rise if we don't reduce carbon emissions quickly, leading to significant impacts on food production (e.g. desertification), water supply (e.g. glacier erosion), ecosystems (e.g. habitat loss), weather patterns (e.g. more floods, storms), and potentially globally destabilizing events (e.g. Atlantic ice

[4] https://earthobservatory.nasa.gov/Features/WorldOfChange/decadaltemp.php

sheet melting) (see HM Treasury 2006). The resulting impacts of these environmental changes would have dramatic social, political, and economic knock-on effects, such as international and civil wars over resource access, droughts and famines from declining water access, mass displacement and movement of people in response to floods and rising sea levels, and much more besides.

Estimates in the Stern Review in the mid-2000s suggested that we need to keep the concentration of atmospheric GHGs at below 550 ppm in order to avoid (hopefully) a 2 °C rise in global temperature and the consequences of global warming (ibid.). The Review expected us to reach this point by 2035. However, more recent estimates in the 2007 Fourth Assessment Report of the IPCC are less optimistic. It's now generally accepted that we need to keep the concentration of atmospheric GHGs to 450 ppm in order to avoid a 2 °C temperature rise, which means "reducing global emissions by up to 85 per cent over 1990 levels by 2050" (Jackson 2009: 12). As a result, the 2015 Paris COP21 agreed to aim for a 1.5° temperature rise, meaning that immediate and significant action needs to be taken by countries around the world in order to reduce their carbon emissions and transition towards low-carbon economies.

Energy and Low-Carbon Transitions

Low-carbon transitions are not easy. Tim Jackson (2009), for example, argues that we've made great strides in improving 'relative decoupling' of growth and emissions—meaning that we're getting more efficient at using resources and energy—in certain countries, but that 'absolute decoupling' hasn't improved. Looking at climate change from a global perspective shows how carbon-embedded our economies remain, since emissions are often hidden by the offshoring of manufacturing. Jackson's conclusion is that the world needs to improve its resource and energy efficiency tenfold within the current carbon-economic parameters, and within the near future (ibid.: 80).

At this point it's important to stress that our economies, our societies, and even our polities have been, continue to be, and will always be defined by carbon (Unruh 2000; Boykoff and Randalls 2009; Bridge 2010; Mitchell 2011). *How* they are defined by carbon is the important question.

Although the effects of releasing fossilized carbon—or fossil energy—have been theorized and considered for some time, attempts to mitigate and ameliorate its impact have only really started in the late-1980s and with the foundation of the UNFCCC in 1992. The main global mechanism for dealing with rising carbon emissions is the Kyoto Protocol which committed its signatory countries to reducing their GHG emissions below 1990 levels over two periods—by five percent between 2008–2012 and by 18 percent between 2013–2020 (Tickell 2009). Attempts to reduce GHG emissions at the national and local scales are generally and necessarily focused on reducing fossil energy production and consumption, which often comes into conflict with other societal priorities (e.g. employment, national income, etc.). Several sectors are key in this regard, including industry, agriculture, power generation, and transport (HM Treasury 2006); all these sectors need to reduce their energy production and consumption.

Using data from the International Energy Agency (IEA), it is possible to map out this energy production and consumption over time, and the following discussion draws on these data from the IEA Headline Global Energy Data.[5] The growth of renewable energy seems impressive in the period covered by the data, from 1971 and 2015. During this time period, for example, the global production of energy from renewables and waste increased from around 700,000 kilo-tonne equivalents (ktoe) to over 1.8 million ktoe—close to 13 percent of total energy production in 2015 versus around 12 percent in 1971. The expansion of renewables, however, has to be positioned in relation to fossil energy production. During the same time period, global production of coal, peat, and oil shale rose from around 1.5 million ktoe to nearly 4 million ktoe—meaning that coal is still a major energy source no matter how much it has been castigated over the preceding decades (Tyfield 2014). Similarly, global production of crude oil, liquid natural gas, and feedstocks rose from about 2.5 million ktoe to nearly 4.5 million ktoe—again, illustrating the extent to which oil still dominates energy geographies (Calvert 2016). It's noticeable how these data support the argument made by Jackson (2009) about a lack of absolute decoupling; that is, although renewable energy production has increased, suggesting rising energy efficiencies, fossil energy production has increased alongside it. Hence why GHG emissions have increased consistently over this time period (see Fig. 3.1 above).

Ending our fossil dependence, or 'carbon lock-in' (Unruh 2000), is key to resolving climate change, but it often seems like an intractable problem. The main solution is—and must be—to transition to a low-

---

[5] https://www.iea.org/newsroom/events/statistics-iea-headline-energy-data.html

carbon economy in which the energy produced and consumed releases (far) fewer GHGs than the current energy regime. What that entails and how to get there are the sticking point. At present, one of the main routes to a low-carbon future is the expansion of bioenergy as a renewable energy source, especially the production of biofuels as a substitute for transport fuels like petroleum/gasoline. With Canada, for example, in 2010 the transport sector represented 28 percent of Canadian GHG emissions, compared with electricity generation at 15 percent and fossil fuel production at 13 percent (Wood and Mabee 2016: 16), meaning that finding an alternative energy source for transport was and is a social, political, and policy priority. Are bioenergy and biofuels a feasible solution to achieve a low-carbon future?

BIOENERGY AND LIQUID BIOFUELS

There are two main reasons I focus on bioenergy and especially biofuels in this book. First, bioenergy represents—both when I started my research in the early 2010s and still today—over half of renewable energy production in major economic jurisdictions like the United States of America (USA) and European Union (EU) (Zimmerer 2011; ClientEarth 2012).[6] Analysing the drivers, policies, and problems of the dominance of bioenergy in these jurisdictions will, therefore, contribute to an understanding of nature-economy relations. Second, in the early 2010s, biofuels were still being touted as a major potential contributor to low-carbon transitions by policy-makers, industry, and others, especially in the USA (Dahmann et al. 2016) and EU (Olsen and Rønne 2016). Finding a substitute for transport fuels is the main goal, since electrification of cars, freight, and airplanes would not provide like-for-like energy substitutions. As a result, many countries positioned bioenergy and biofuels as a major plank in their low-carbon transition strategies—for example, Australia, Germany, Sweden, the UK, and the USA (Trottiers 2013). Understanding how bioenergy and biofuels came to be viewed as a major plank in low-carbon transitions, and then enacted as such, opens up analytical space to examine the co-construction of markets and natures.

[6] https://www.eia.gov/totalenergy/data/monthly/index.php#renewable

Bioenergy: From Old to New

As noted above, bioenergy is seen as an important contributor to reaching a low-carbon future. So what is it? And how might it contribute to low-carbon futures? Starting with the first question raises an immediate peculiarity. Humans have been using 'biomass' (i.e. biological matter like plants, wood, and waste) as an energy source—hence 'bio'energy—for millennia, whether burning wood or peat to generate heat or eating plants to generate bodily energy.[7] Across the world today, biomass remains a major energy source, especially in the Global South, although many people associate modern living standards with the shift to fossil energy (Calvert et al. 2017). At the same time, however, bioenergy is increasingly touted as a potential renewable energy source in the Global North, where it is associated with things like wood pellets as substitutes for coal in power stations, or biofuels as substitutes for petroleum/gasoline in transport fuels. As I discuss below, and in other chapters in this book, there is a growing range of policy visions and strategies promoting the expansion of a 'bio-based economy'—or 'bio-economy'—as a possible low-carbon transition pathway based on the idea of replacing fossil energy with bioenergy. Whether the planet can grow enough biomass to support this transition is a critical question, and probably has to be answered in the negative (Pimentel and Patzek 2006; Pimentel 2009), a point I come back to below.

The potential of bioenergy—and the bio-economy more generally—to contribute to low-carbon transitions lies in the diverse applicable uses of biomass across our political economies *and* the notion that renewable resources (e.g. plants), at least in our lifetimes, are more sustainable, by definition, than non-renewable resources (e.g. oil). As mentioned already, biomass is an energy source, but it's also an (environmental) input into many other political-economic processes—especially as we develop more sophisticated biotechnologies (Meijl et al. 2015). As a long-standing 'natural' resource (Bridge 2009), biomass like plants, trees, and organic waste contributes to more traditional sectors like food production, papermaking, and construction. More recent biotechnological processes have opened up an array of other uses for biomass, such as the production of chemicals, materials, and liquid fuels—biomass is even touted as the 'new

---

[7] The EU's definition of 'biomass' is "the biodegradable fraction of products, waste and residues from biological origin from agriculture (including vegetal and animal substances), forestry and related industries including fisheries and aquaculture, as well as the biodegradable fraction of industrial and municipal waste" (quoted in Olsen and Rønne 2016: 169).

oil' (Shortall et al. 2015). The focus on using biomass to produce energy has a number of social and policy drivers, which I discuss below in reference to biofuels, but is most obviously associated with finding new ways to replace fossil energy. As Table 3.1 illustrates, bioenergy is premised on the technological conversion of biomass into energy substitutes for fossil fuels (e.g. oil, coal, etc.). Electricity, for example, can be generated from the burning of biomass, usually densified as pellets or briquettes, rather than coal or natural gas (Calvert et al. 2017).

While there are material limits to the use and usefulness of bioenergy (Birch and Calvert 2015), it's still a major renewable energy source in jurisdictions like the USA and EU, as mentioned already (Zimmerer 2011; ClientEarth 2012). In part, this is because bioenergy can (seemingly) substitute for fossil energy socially, economically, and materially with limited need for institutional or infrastructural change—or, at least, that's the per-

Table 3.1   Bioenergy pathways

Bioenergy	Technology	Biomass	Substitution
Ethanol	Fermentation corn or wheat	Cereal, grain, and fodder crops	Petroleum/ gasoline
	Cellulosic: (1) Hydrolysis (chemical/ enzymatic) (2) Fermentation (3) Gasification	Cellulosic residues (forest, agriculture, municipal waste)	Petroleum/ gasoline
	+ cogeneration of heat and power	Cellulosic residues (forest, agriculture, municipal waste)	Natural gas Coal
Biodiesel	Transesterification: (1) Oilseed (2) Tallow (3) Yellow grease	Oilseeds, canola, soybean, flaxseed, tallow, yellow grease	Diesel
	Cellulosic gasification to Fischer-Tropsch fuels	Cellulosic residues (forest, agriculture, municipal waste)	Diesel
	+ cogeneration of heat and power	Cellulosic residues (forest, agriculture, municipal waste)	Natural gas Coal Nuclear
Bioelectricity	Cogeneration of heat and power	Cellulosic residues (forest, agriculture, municipal waste)	Natural gas Coal Nuclear

Source: adapted from Wood and Mabee (2016: 58)

ception and expectation (see Chaps. 4 and 5). As such, it represents an ideal market-based—even 'neoliberal'—transition pathway; that is, all we seem to need to do is ensure that bioenergy is competitive with fossil energy, perhaps through the introduction of carbon pricing (Felli 2014), and the market will sort itself out (see Scarlat et al. 2015). Framing bioenergy this way is politically attractive because it means there is no need for major societal change, in that our lifestyles, landscapes, institutions, and infrastructures can stay the same—basically, (environmental) transition without (economic) disruption. Obviously, things are more complicated than this, as will become evident throughout this book. One example of these complexities is the need to adapt existing value chains in order to redirect or create anew biomass supply chains, reflecting the biophysical materialities of biomass and its potential use in the bio-economy (e.g. seasonal and annual variability, storage needs, pre-treatment to reduce moisture, densification to reduce bulk, spatial distribution, etc.) (Yue et al. 2014; Birch and Calvert 2015).

Understanding Liquid Biofuels

Liquid biofuels are one of the most attractive bioenergy pathways and have been the centre of policy attention in North America and the European Union (EU) since at least the early 2000s—other jurisdictions like Brazil have even longer policy histories (Le Bouthillier et al. 2016). Biofuels have been around a long time, though, stretching back to the early twentieth century. As James Smith (2010: 15) notes, biofuels are fairly simple 'technologies', referring to the "energy derived from biomass through processes such as combustion, gasification or fermentation". Generally, biofuels are split between 'generations' depending on the feedstock, conversion technologies, and sustainability criteria they meet; for example, most people differentiate between 'first', 'second', and 'third' generation biofuels, as well as between 'conventional' and 'advanced' biofuels (Charles et al. 2007; Dahmann et al. 2016: fn52).

Conventional Biofuels

First-generation biofuels—or 'conventional'—usually refer to ethanol produced from food crops with high levels of sugar or starch content (e.g. sugarcane, corn, wheat), as well as to biodiesel produced from oilseed crops (e.g. canola, soybean, palm, *Jatropha*, etc.). Ethanol represents the most commonly produced first-generation biofuels, with about 1,614,000 barrels/day produced globally in 2014, compared with 529,000 barrels/day of biodiesel.[8] Certain countries and jurisdictions dominate this ethanol and biodiesel production. According to data from the USA's Energy Information Administration (EIA), for example, as of 2014 the top six ethanol producers around the world represent 96 percent of global ethanol production (see Fig. 3.2, left graph). Similarly, the top six biodiesel producers represent 90 percent of global biodiesel production (see Fig. 3.2, right graph).

The main ethanol producers are the USA (58 percent) and Brazil (27 percent), by some way, although the former was behind the latter until 2006. As Fig. 3.2 shows, the USA has come to dominate ethanol production in a relatively short space of time, essentially since the 2005 Energy Policy Act established the first Renewable Fuel Standard (RFS) and the 2007 Energy Independence and Security Act created the second RFS (Dahmann et al. 2016). I will come back to these below, so I will not dwell on them here. Although Canada is in the top six ethanol producers, it only produces 1.9 percent of global ethanol (in 2014). In relation to biodiesel, the main producer has always been the EU, which now represents 39 percent of the global production. Other countries, however, are increasing their production levels, especially in the Global South (e.g. Brazil, Indonesia, Argentina, and Thailand).

As the main conventional biofuel, ethanol holds an ambiguous place in assessments of the potential of bioenergy as a contributor to low-carbon transitions for several reasons—leaving aside other issues like land-grabbing, terrible work conditions, and so on (Shiva 2008; Bello 2009; Smith 2010; Backhouse 2015).

First, and most importantly, two influential scientific studies came out in 2008 which suggested that the effects of indirect land-use change (ILUC)—that is, the effects of using agricultural land for biofuels on non-agricultural land, notably the increase in land clearing—meant that the promised GHG emissions reductions from ethanol were actually a mirage (Fargione et al. 2008; Searchinger et al. 2008). Searchinger et al. specifically claimed that certain types of ethanol production—primarily corn ethanol in the USA—were worse than fossil energy use, thereby countering the idea that biofuels are by definition a sustainable and low-carbon transport energy. Others, like Pimentel (2009: 207), claim that it took "146% more energy … to produce a gallon of corn ethanol than is in the

[8] https://www.eia.gov/beta/international/data/browser/index.cfm

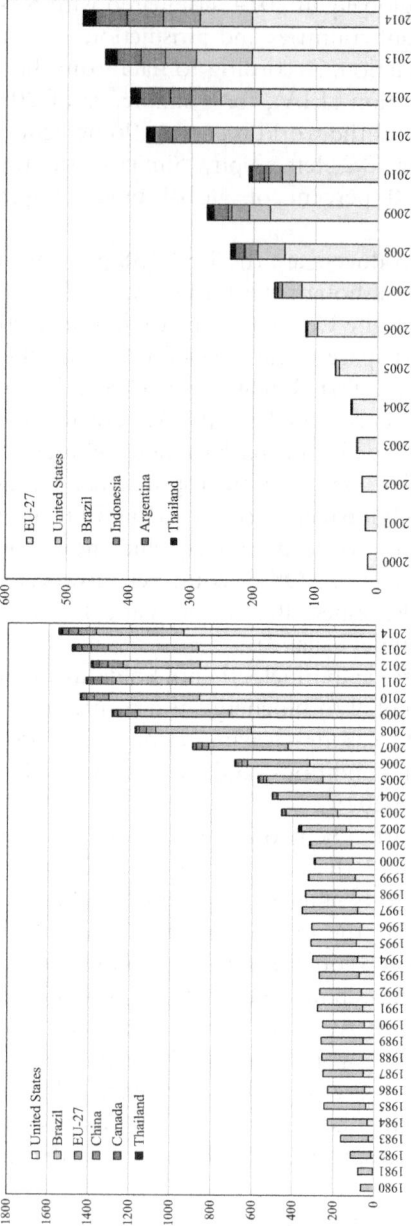

Fig. 3.2 Ethanol (left) and biodiesel (right) production, top six countries, 1000 barrels/day. (Source: US Energy Information Administration, International Energy Statistics)

ethanol itself". Second, alongside these scientific articles, a 2008 World Bank report also linked increased ethanol production to rising food prices (Mitchell 2008). A number of scholars have made related arguments. For example, McMichael (2009: 826) argues that ethanol has opened up "new profit frontiers for agribusiness, energy and biotechnology corporations" and shifted responsibility for emissions reductions to the Global South. Similarly, Gillon (2010) argues that ethanol production, specifically in the USA, has led to the financialization of agriculture through rising (venture and institutional) investment in long-term commodity contracts. The result, in both cases, is rising food prices.

Third, data collected and analysed by Baines (2015) shows that ethanol production is correlated with (a) other uses, like animal feed (negatively), and (b) oil prices (positively). It's interesting to note that around 1980 corn and oil prices became correlated, shortly after the USA's 1978 National Energy Act which introduced a subsidy for blended petroleum/gasoline (Dahmann et al. 2016). As such, it is unclear how biofuels will provide an alternative to fossil fuels when they're so dependent on the latter. Fourth, there are a range of other concerns with ethanol ranging from its impact on environments and biodiversity, considering the need to expand agricultural monocultures to expand corn production and the increased use of pesticides and fertilisers this would entail (Hill et al. 2006). Finally, there are simply not enough food crop feedstocks to meet petroleum/gasoline demand in countries like the USA; for example, as Pimentel and Patzek (2006: 875) put it, "the 18 percent of the US corn crop that is now converted into 4.5 billion gallons of ethanol replaces only 1 percent of US petroleum consumption" (in the early 2000s). A more basic problem is that current biofuels production only meets two to three percent of world transport fuel demands, implying that substituting for oil would require a significant increase in land devoted to feedstock cultivation (Smith 2010; IEA 2017).

Advanced Biofuels
As a result of rising environmental, economic, and social concerns about first-generation biofuels, there's been a growing push behind second (and third)-generation biofuels—usually called 'advanced biofuels'. The primary difference between first- and second-generation biofuels is the feedstock and conversion technology platform used to turn that feedstock into a liquid fuel (Sims et al. 2010). The different biofuel pathways, including conventional ethanol and biodiesel, are illustrated in Fig. 3.3. Second-generation

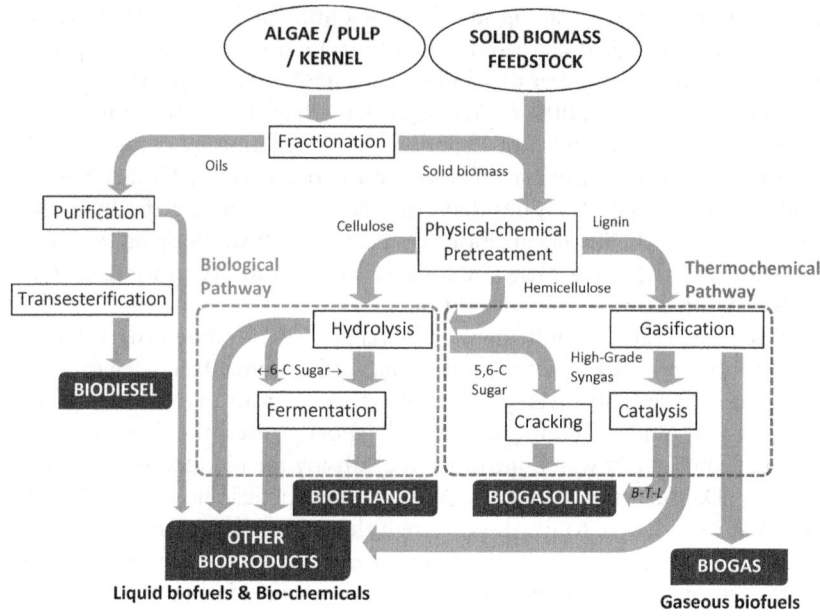

Fig. 3.3 Biofuel technology pathways. (Source: Warren Mabee, Queen's University, 2018, reproduced with permission)

biofuels are primarily derived from lignocellulose feedstocks like non-food crops (e.g. wood), energy crops (e.g. *Miscanthus*), agricultural and forest residues, and municipal organic waste (Worldwatch Institute 2007). Second-generation biofuels are often called 'cellulosic biofuels' because they primarily entail the conversion of lignocellulose into liquid fuels. While there are several technological conversion routes, the main two are bio-chemical and thermochemical approaches (Sims et al. 2010; Calvert et al. 2017). Biochemical approaches include acid and enzymatic hydrolysis, which involve the breaking down of biomass into sugars using acid or enzymes before its fermentation into ethanol. Thermochemical approaches include gasification and pyrolysis, which involve the conversion of biomass into a synthesis gas ('syngas') that can be catalysed to form biofuels. While there are interesting back stories and applications of all these conversion technologies, the key issue today is whether they're economical.

Different biomass has different biophysical qualities—that is, biophysical materialities—which means that there is a range of different potential feedstocks for advanced biofuels production. It is technologically possible to turn almost any biomass into a liquid or gas fuel; the issue is whether it can be converted into a fuel at an economical price (Smith 2010). A biofuel that costs US$200 per barrel is no use to anyone at present, even if it might represent an amazing technological breakthrough. Much of the focus on advanced biofuels has, therefore, centred on increasing the efficiency of conversion technologies, expanding feedstock sources, and scaling up production facilities (e.g. Worldwatch Institute 2007; Sims et al. 2010; IEA 2011; Stephen et al. 2011). Evidently the technical development of advanced biofuels is driven by specific techno-economic concerns with things like technology uncertainties (e.g. conversion rates, enzyme costs, etc.), capital costs required to build large (bio-)refineries and create the necessary feedstock supply chains, and poor economic or environmental performance of new and existing biofuels companies (Stephen et al. 2011). It's notable, however, that there has been significant policy and political support—sometimes from across the political spectrum—behind advanced biofuels.

Biofuels Policy Regimes and Policy Drivers

Different jurisdictions have very different policies in place to support both conventional and, increasingly, advanced biofuels, many of which have been put in place over the last few decades. It's useful to outline the policies enacted in a few key countries in order to show what has driven the implementation of these different policies and how they've configured the resulting biofuels 'market'. Considering the information in Fig. 3.2 above, I'm going to discuss the policy context and implementation in Brazil, the USA, and the EU—the three main producers—and leave discussion of Canada, my primary research focus, to later chapters.

Brazil
Brazil was one of the earliest promoters of biofuels, especially ethanol, and its policy regime was driven primarily by recurrent energy crises and an abundance of sugarcane as a potentially more secure alternative. The 1973 oil crisis put an enormous pressure on Brazil's ability to balance its imports and exports—according to Smith (2010: 22), for example, oil imports *before* the crisis represented half of Brazil's exports, so afterwards they rose even further. As a result, the government established the Proálcool programme in 1975, which consisted of tax incentives and blending mandates, leading to investment in and expansion of the ethanol sector (Worldwatch Institute

2007). Further government support for institutional and infrastructural changes included government support for ethanol-ready automobiles as well as the roll-out of ethanol distribution (e.g. blended petrol pumps), consumer incentives, and grants and loans to sugar and ethanol producers (Smith 2010; HLPE 2013). Even though these market development policies—see Chap. 6 for a look at Canada—have been gradually withdrawn, blended ethanol is now thoroughly embedded as a major (bio)energy source. No doubt, the fact that Brazil has a blending mandate of 27 percent of petroleum/gasoline (raised in 2015) helps, although the exact mandate has waxed and waned over time between 20 and 27 percent since 1993 (Rico 2008; HLPE 2013). As a result of these policies, Brazil became the world's leading ethanol producer up until 2006 when the USA superseded it. Brazil's ethanol production has grown from 64,000 barrels/day in 1980—when it was the basically the world's only ethanol producer—to 195,000 b/d in 1993, when a 20 percent plus mandates came in, up to 430,000 b/d in 2014.[9] There are downsides to all of this, of course, including the deforestation of the Amazon, land-grabbing, poor working conditions, and land evictions leading to landlessness (Smith 2010; Backhouse 2015). It has also likely stymied the development of advanced biofuels in Brazil, since the policy drivers of rural development, emissions reductions, and energy security are largely achieved within a sugar-based ethanol regime.

United States of America
The world's largest ethanol producer today is the USA, primarily as the result of policies designed to convert corn into ethanol. Early policy supports included subsidies for ethanol production introduced with the 1978 Energy Policy Act. Despite this, US ethanol production did not rise above 100,000 b/d until 2000 and only surpassed Brazilian production levels after concerted policy efforts in the mid-2000s.[10] A suite of policies help explain the huge expansion of US ethanol production, including: the 2000 Biomass R&D Act, which was meant to support the development of bio-based energy and products; the 2002 Farm Security and Rural Investment Act (Farm Bill), which supported biofuels through grants, R&D, and procurement rules; the 2005 Energy Policy Act, which introduced a Renewable Fuel Standard (or RFS1) for the first time alongside tax incentives, loan guarantees, R&D programmes, and suchlike; and the 2007

[9] https://www.eia.gov/beta/international/data/browser/index.cfm
[10] https://www.eia.gov/beta/international/data/browser/index.cfm

Energy Independence and Security Act, which introduced RFS2 (Power 2016). The main planks of the USA's ethanol programme are RFS1 (2005–2007) and RFS2 (post-2008). These standards set volume requirements for the production of biofuels, both conventional and advanced, and differentiate between conventional and advanced as well as between different types of feedstocks and the GHG emissions required of each (Tyner 2010a).

According to Schnepf and Yacobucci (2013), RFS1 introduced a volumetric mandate that required 'obligated parties'—basically oil refiners, blenders, and importers—to add four billion gallons of biofuels (primarily ethanol at this point) to their petroleum fuels starting in 2006 with the aim of blending 7.5 billion gallons by 2012. With the introduction of RFS2 in 2007, however, these mandates changed to nine billion in 2008 and 36 billion gallons by 2022. Moreover, by 2022, the amount of corn ethanol was supposed to be capped at 15 billion gallons with 16 billion gallons coming from cellulosic biofuels (ibid.). Other changes introduced by RFS2 included stricter GHG, land-use, and biomass requirements for biofuels, although the intentions behind these changes were largely undone by the decision to exempt existing facilities from these requirements (Power 2016). As a result of these changes, the USA has become the world's main biofuels producer, although the initial surge in production—rising nearly 600 percent between 2003 and 2010—has since largely plateaued, sitting at around 15 percent from 2010 to 2017.[11] Despite attempts to stimulate advanced biofuels within the RFS system, the operations of the RFS have created significant uncertainty for the development of advanced biofuels, and the cellulosic mandate has been reduced consistently as a result (Schnepf and Yacobucci 2013). On the one hand, the Environmental Protection Agency (EPA) sets annual targets for cellulosic biofuels, which are contested by incumbents and hard to predict; and on the other hand, the RFS have exempted facilities built before 2009 from the GHG emissions requirements, leaving whole swathes of the ethanol industry free to operate without pressure to improve sustainability (Tyner 2010b; Power 2016).

European Union
Finally, the European Union (EU) has become a major player in the biofuels arena, especially when it comes to biodiesel (see Fig. 3.2, right hand) but also with regard to the establishment of (global) sustainability criteria.

[11] https://www.eia.gov/beta/international/data/browser/index.cfm

Early support for biofuels came through the Common Agricultural Policy (CAP); for example, in the early 1990s the CAP introduced an oilseed crop quota, tax exemptions, and non-food set-aside land requirements, which allowed the cultivation of energy crops (Londo and Deuwaarder 2007; Sorda et al. 2010). Primarily driven by concerns about rural development and agricultural surpluses, the EU's policy framework reoriented around climate change in the early 2000s following the 2001 Renewables Directive, which set national renewable energy targets for member states (Olsen and Rønne 2016). Policies in this period included the 2003 Biofuels Directive, which set a two percent biofuel target in the transport sector by 2005 and 5.75 percent by 2010, although it was not necessarily mandatory for member states; the 2005 Biomass Action Plan, which aimed to promote bioenergy more generally; the 2006 Strategy for Biofuels, which sought to promote biofuels in the EU and Global South; and the 2009 Renewable Energy Directive (RED), which sought to integrate sustainability concerns about conventional biofuels into policy supports for both conventional and advanced biofuels (Londo and Deuwaarder 2007; Worldwatch Institute 2007; Sorda et al. 2010; Olsen and Rønne 2016). With RED, the EU set minimum biofuel target of ten percent of transport fuels by 2020, as part of a broader package around renewable energy, although only for biofuels that reduce GHG emissions by 35 percent compared with fossil fuels (Sorda et al. 2010). Subsequent policy agreements have extended the renewable energy target to 27 percent of energy consumption by 2030 (Olsen and Rønne 2016).

There are several differences between the USA and EU worth noting. Most obvious, as illustrated in Fig. 3.2, is the fact that the USA is dominated by ethanol biofuels, while the EU is dominated by biodiesel. As a result, the EU has become reliant on oilseed imports to reach its mandate (HLPE 2013), which might explain why it has played a significant role in promoting global sustainability criteria (Bailis and Baka 2011; Ponte 2014). Other differences resulting from this are worth noting, however. First, and in contrast to the USA, the EU's approach has been to set proportional mandates (cf. volumetric ones)—that is, mandates for a certain percentage of transport fuels to come from biofuels. As Tyner (2010a) notes, volumetric targets have created a corn-ethanol bias in the USA, since this biofuel type is needed to meet the volume requirements. As a result, the US system penalizes certain technology conversion pathways (e.g. thermochemical), while the EU system leaves the conversion pathways open for producers. Second, the EU is largely feedstock and technology 'neutral' in that its standards and regulations don't differentiate between them, with all biofuels needing to meet specific sustainability criteria after RED—starting at 35 percent GHG emissions reductions and rising to 50 percent in 2017 and 60 percent in 2018 (ClientEarth 2012). Third, in 2012, the EU sought to introduce a cap on the use of food crop-based biofuels, while retaining the overall mandate, thereby necessitating a significant increase in advanced biofuels production (HLPE 2013). Generally, the EU seems to offer a more open context in which to develop advanced biofuels, whereas US policy was and still is embedded in the political-economic expectations around corn production and agricultural politics.

Where Are Advanced Biofuels Today?

Anyone reading this book could quite legitimately ask at this point, "where are all these biofuels we were promised?" Although both the USA and EU have sought to promote advanced biofuels in their own ways (see Balan et al. 2013), the success of their policies is equally ambiguous—and some people would argue that advanced biofuels remain as much of a pipe dream today as they were in the early 2010s when I first started researching them. Despite attempts to stimulate and create a market of and for advanced biofuels—which I look at in more depth in Chap. 6 when it comes to Canada—nowhere have cellulosic biofuels become a significant proportion of total biofuels, let alone general transport fuels. For example, the original American RFS volume requirements for 2014–2019 stipulated three billion gallons of cellulosic biofuels in 2015, 4.25 billion in 2016, and 5.5 billion by 2017; however, the EPA had to reduce these mandates to 0.123 billion, 0.230 billion, and 0.311 billion, respectively, as cellulosic production could not match expectations (Irwin and Good 2017). Similarly, EU production of cellulosic biofuel has been limited, only reaching 0.24 billion litres in 2016—or 0.063 billion gallons (ePure 2017). Across these jurisdictions, a number of common problems with the expansion of advanced biofuels are surfacing, especially as this relates to their potential contribution to low-carbon futures. As a result, liquid biofuels generally—both conventional and advanced—are increasingly framed within a broader agenda focused on the so-called bio-based economy or bio-economy. And it is to this broader agenda that I turn next.

The Emerging Bio-Economy

In November 2015, over 700 people from 80 countries gathered in Berlin for the first Global Bioeconomy Summit organized by the German Bioeconomy Council—or Biöoekonomierat.[12] A new advisory body set up by the German Federal Government, its website states that the Bioeconomy Council is meant to advise "the Federal Government on the implementation of the 'National Research Strategy Bioeconomy 2030' and the 'National Policy Strategy on Bioeconomy'".[13] Its aims are to create the "optimum economic and political framework conditions for a biobased economy", also according to its website, and it has council members drawn from academia, business, and non-government organizations. According to the Council, the bio-economy it has been set up to support is defined as "the production and utilization of biological resources (including knowledge) to provide products, processes and services in all sectors of trade and industry within the framework of a sustainable economy". Combining 'nature's cycles' with 'research for innovation', the bio-economy will help 'green transitions' and help 'solve global problems'. A post-2015 conference statement by members of the Bioeconomy Council, printed in the science journal *Nature*, argued that the bio-economy will contribute to achieving the Sustainable Development Goals (SDGs), including, for example, Energy for All (SDG7), Sustainable Cities (SDG11), and Combat Climate Change (SDG13) (El-Chichakli et al. 2016). The Bioeconomy Council hosted a second Global Summit in 2018.

What Is the Bio-Economy?

I began studying the bio-economy in the mid-2000s (Birch 2006), just as it was emerging as a key concept in policy documents produced by the Organisation for Economic Co-operation and Development (OECD) and European Commission (EC) (see CEC 2005; OECD 2005). The fact that there are entities like the Bioeconomy Council today, and Global Summits organized around the topic, speaks to the journey that the concept and its

---

[12] http://biooekonomierat.de/en/bioeconomy/
[13] An interesting connection can be made here between ordoliberalism—a variant of neoliberalism according to many (e.g. Foucault 2008; Peck 2010; Birch 2017)—and the (stateled) construction of 'framework conditions' for markets, that is, the conditions necessary to ensure market competition, to reduce potential monopolies, and to support wider 'social' goals (e.g. employment, social stability, etc.).

policy impact have had since then. From a rather obscure, jargonistic term, the 'bio-economy' has spread globally until numerous countries, in the Global North and South, now have dedicated bio-economy policy visions and strategies, or bio-economy-related strategies (Bioeconomy Council 2015a, b). See Fig. 3.4 produced by the Bioeconomy Council for examples. I've already outlined the Bioeconomy Council's definition of the bio-economy, but this definition is by no means definitive, nor is it uncontested. Recently, a range of literature has engaged with the bio-economy as a concept (analytical definition), particularly its deployment by policymakers in their future visions (substantive definition) of low-carbon transitions. I start this discussion with the substantive definitions by drawing on several reviews of supra-national, national, and regional bio-economy strategies (e.g. McCormick and Kautto 2013; Staffas et al. 2013; Pfau et al. 2014; Bugge et al. 2016; Hausknost et al. 2017; Birner 2018). After that, I turn to the analytical definitions, primarily drawing on my own and others' existing research.

A simple definition of the bio-economy is the conversion of biomass into energy, products, and services, alongside and distinct from food production. However, the bio-economy is framed in different ways in different policy strategies, meaning that it can be a confusing concept to use. It first appears in a coherent form in documents produced by the OECD and EC in the mid-2000s, although each entity has different rationales. In a 2004 report titled *Biotechnology for Sustainable Growth and Development*, the OECD emphasized the important role the 'bio-based economy' could have in decoupling economic growth from ecological impacts; they defined the 'bio-based economy' as "an economy that uses renewable bioresources, efficient bioprocesses and eco-industrial clusters to produce sustainable bioproducts, jobs and income" (OECD 2004: 5, fn1). A year later, in 2005, the OECD started to develop a 'bio-economy' policy agenda designed for government implementation. Here, the OECD (2006: 1) defined the 'bio-economy' as:

> the aggregate set of economic operations in a society that use the latent value incumbent in biological products and processes to capture new growth and welfare benefits for citizens and nations.

Around the same time, according to Chris Patermann—who is the former Programme Director of Biotechnology, Agriculture & Food Research at DG Research—the EC began to consider the potential of 'white biotechnology',

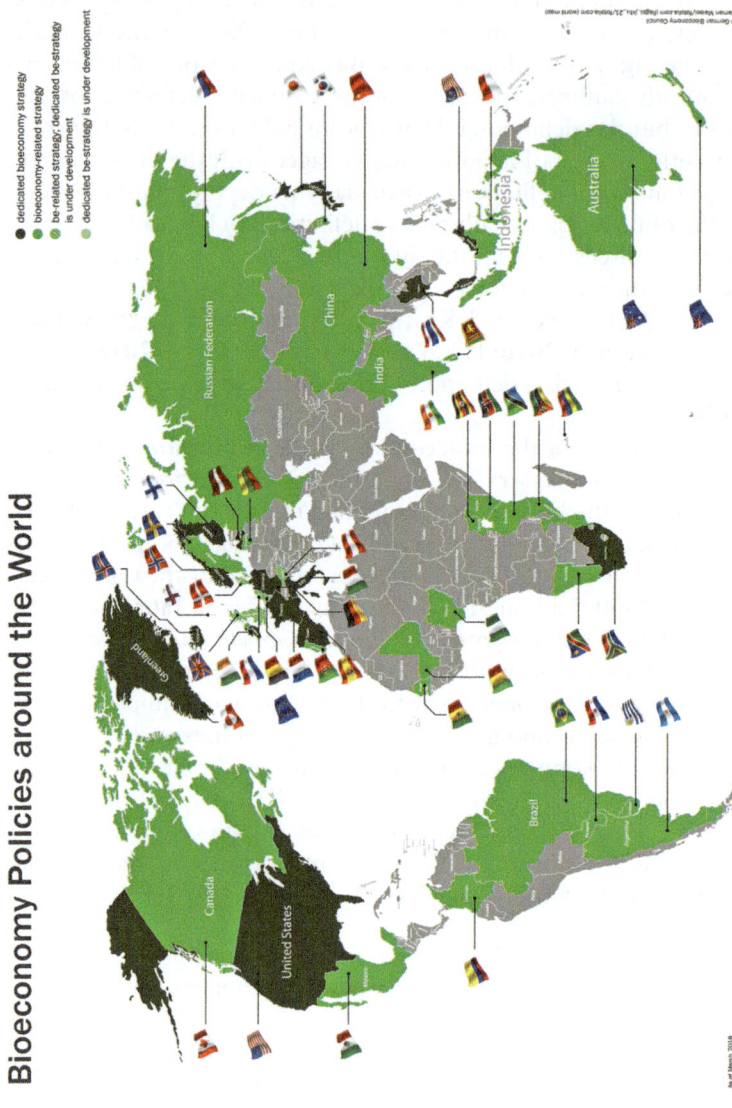

Fig. 3.4 Bio-economy strategies around the world. (Source: German Bioeconomy Council 2018, reproduced with permission)

centred on industrial applications, for the European economy, especially within a public context hostile to the agricultural use of modern biotechnology (e.g. genetically modified organisms, or GMOs) (Bonaccorso 2015). At a 2005 EC Conference, people like Patermann and his boss Janez Potočnik—the former Science and Research Commissioner—presented the idea of a 'knowledge-based bio-economy', or KBBE, as a way to refer to "the sustainable, eco-efficient, transformation of renewable biological resources into health, food, energy and other industrial products" (CEC 2005: 3). From this starting point, the KBBE expanded as a concept underpinning research funding, primarily through the EU's Framework Programme (FP). Today, over ten years later, the DG Research and Innovation of the EC has its own Bioeconomy Directorate and the EC has its own dedicated bio-economy strategy titled *Innovating for Sustainable Growth: A Bioeconomy for Europe* (CEC 2012).

Since the mid-2000s, a number of other countries and jurisdictions have developed their own bio-economy strategies. Some of the most thorough and detailed analyses of these policy strategies have been produced by Kes McCormick and colleagues (e.g. McCormick and Kautto 2013; Staffas et al. 2013; McCormick 2014; de Besi and McCormick 2015). In their analysis of the European bioeconomy, McCormick and Kautto (2013) and McCormick (2014) highlight the difference between the OECD (2006, 2009), EC (CEC 2005, 2012), and US (White House 2012) strategies and definitions. The OECD strategy is primarily centred on the development of bio-based products, while the EC strategy is both product-centred as well as placing emphasis on the efficient use of renewable and natural resources. In contrast, the US strategy is more centred on (biotechnological) research and innovation, including the biomedical sector. Most importantly, perhaps, the EC strategy is more oriented towards the idea of transitioning society towards a low-carbon future. In their examination of other national strategies, Staffas et al. (2013) argue that different strategies use different terms, distinguishing in particular between 'bio-based economy' (BBE) and 'bio-economy' (BE). Policy strategies using the former definition focus on the natural resource base (e.g. biomass), rather than the technological conversion process (e.g. biotechnology), which is the focus of the latter definition. Countries and entities like the EC, Germany, Finland, and Sweden use the BBE definition, while the OECD, the USA, Canada, and Australia use the BE definition (ibid.). It is possible to see a simple association between the BBE notion and social democratic political economies, and between the BE notion and (neo) liberal political economies—implying that more market-centred countries are less inclined to wholesale transition. Finally, de Besi and McCormick (2015) delve into regional and industrial strategies, alongside national ones. Many of these strategies are more focused on specific forms of market development policies needed to create the bio-economy, including procurement requirements, regional networking platforms, public acceptance policies, and so on. I come back to the specifics of the Canadian example in Chaps. 4, 5, and 6.

On the analytical side, there are several different ways to define the bio-economy as well—I cannot go through all of them, considering the burgeoning writing in this area, so I will focus on a few key examples. One early perspective by Frow et al. (2009) identifies the different 'politics of plants', covering topics from climate change to economic growth. In their view, the bio-economy is a system in which competing visions and pressures (will) drive the specific goals and activities. Richardson (2012) and Schmid et al. (2012) represent examples of other early attempts to parse the different uses of bio-economy in policy discourse. First, Richardson argues that the bio-economy entails two strategies, one of 'appropriationism' and the other of 'substitutionism'; the former entails taking control of natural resources (e.g. land, plants, etc.), while the latter entails replacing fossil materials with plant-derived ones. Like Frow et al., Richardson argues that the bio-economy's attractiveness is derived from the fact that it can be 'something for everyone'—farmers, environmentalists, politicians, business, and so on. Second, Schmid et al. argue that the EC and OECD focus on biomass *and* biotechnology as the key identifiers of the concept, which narrows the concept considerably and downplays alternative bio-knowledges like agroecology (also Kitchen and Marsden 2011; Levidow et al. 2013). In their analysis of the literature on the bio-economy and sustainability, Pfau et al. (2014) highlight some of the contentious issues around the bio-economy raised in Frow et al., Richardson, and Schmid et al. In particular, Pfau et al. differentiate between academic literatures that (1) assumes an inherent sustainability benefit to the bio-economy (e.g. renewable), (2) posits a conditional sustainability benefit (e.g. resource efficiency), (3) criticizes these sustainability claims (e.g. land competition), and (4) highlights the significant problems with sustainability claims (e.g. invasive species). From a more philosophical perspective, Zwart et al.'s (2015) review of the ethics of a 'bio-based society' places an emphasis on the circularity and renewability inherent in the bio-economy, as well as naturalization of biophysical processes in commercial developments (e.g. 'nature as factory' or 'plants as oil wells'). The bio-economy can be presented as 'natural' as much as it is 'social' (or 'economic') in this regard.

It is notable, across these debates, that there is no singular conception of the bio-economy at play in policy or scholarly circles, something I illustrate in Chap. 4 when discussing Canadian policy imaginaries. In their systematic review of the burgeoning academic literature on the bio-economy, for example, Bugge et al. (2016) identify three prevalent, yet different, visions of the bio-economy in academic concepts: (1) biotechnology vision, (2) bioresource vision, and (3) bioecology vision. The first of these reflects the increasing emphasis on biotechnology as a driver of industry and fount of economic growth (e.g. Kircher 2012); the second reflects the interest in natural resources and their use to produce bio-based products and services (e.g. Matthews 2009); and the third reflects the importance placed on ecological processes in relation to broader concerns with environmental degradation (e.g. soil quality, biodiversity loss, monocultures, fertiliser usage, etc.) (e.g. Kitchen and Marsden 2011). Similarly, Mukhtarov et al. (2017) identify three largely academic debates around the 'neoliberalization' of the bio-economy (e.g. Birch 2006), the sustainability benefits of the bio-economy (e.g. Ostergard et al. 2010), and the societal embedding of the bio-economy (e.g. Goven and Pavone 2015).

From my own perspective, the bio-economy is probably best thought of—analytically speaking at least—as a grand or master narrative or discourse that is embedded within, being both informed by and informing in turn, a specific policy framework (Birch et al. 2010, 2014; Levidow et al. 2012, 2013; Ponte and Birch 2014; Birch 2016a, b; Hausknost et al. 2017). Others have made similar arguments, including Hilgartner who describes the OECD's bio-economy agenda as 'anticipatory machinery' (Hilgartner 2007) and the visions of synthetic biology as 'vanguard imaginaries' (Hilgartner 2015). With these analyses, Hilgartner is getting at the performative character of these future visions and narratives. That is, they're not primarily describing something 'out there'; rather, they're advancing a particular future as desirable and others as undesirable, thereby shaping the preferences and decisions of social actors to bring about those desired ends (also Ponte and Birch 2014). In light of these sorts of argument, and my own analyses, I therefore tend to conceptualize the bio-economy as:

...a concept that refers to the sustainable use of biological, renewable materials in the development of bio-based products, services and energy that substitute for existing fossil fuel-based products, services and energy, as part of a broader societal transition to a low-carbon future. (Birch 2016a: 15)

My definition tries to combine a vision of the process underpinning the bio-economy (e.g. "sustainable use of biological, renewable material") with its outcomes (e.g. "the development of bio-based products, services and energy"), how these outcomes will impact us now (e.g. "that substitute for existing fossil fuel-based products, services and energy"), and why this actually matters (e.g. "part of a broader societal transition to a low-carbon future"). Moreover, I try to avoid the suggestion that the bio-economy is either the solution to everything or the only transition pathway that we can adopt to solve the environmental (and societal) problems we face in the twenty-first century.

CONCLUSION

In this chapter, I've sought to provide a necessarily brief background to the emergence of the bio-economy, especially in relation to liquid biofuels. I discussed climate change at the start in order to set the stage for understanding where the policy rationale and drivers for supporting and promoting biofuels and the bio-economy came from. Although the route has been rather circuitous, the policy rationale and drivers have ended up—generally speaking—centred on the search for socio-technical pathways that can get us to a low-carbon future. Biofuels are one such pathway, embedded within a broader shift from a fossil fuel-based economy to a bio-based economy, or 'bio-economy'. It's important, however, when trying to understand the emergence of *the* bio-economy to acknowledge that there are varied and different pathways to the bio-economy. This is illustrated by the already evident and very different biofuel regimes around the world: for example, Brazil's sugarcane-driven regime, America's corn-driven regime, and the EU's biodiesel-driven regime. All of this discussion helps to frame the bio-economy as a potential solution to a range of environmental problems, but it leaves open what the bio-economy actually means—how to define it—and what it means for different countries, such as Canada.

Where the substantive and analytical definitions I discussed above meet is when it comes to considerations of the socio-technical transformation implied by the bio-economy. Although the policy literature is, perhaps, a little more coy in this regard—not wanting to put politicians or publics off the idea—there's still a clear attempt to paint the bio-economy as a broad-based social, technical, and ecological change in our societies and economies. Where a lot of them differ, though, is the extent to which this societal transition is characterized as a wholesale restructuring or renovation of the prevailing economic system—characterized by many people as (rampant or not so rampant) neoliberalism. How this societal transition is envisaged or imagined, and how those visions are being implemented—usually partially right now—is an empirical question I aim to address next, focusing on Canada as my particular case study although drawing on insights from other parts of the world.

References

Backhouse, M. (2015) Green grabbing – The case of palm oil expansion in so-called degraded areas in the eastern Brazilian Amazon, in K. Dietz, B. Engels, O. Pye and A. Brunnengraber (eds) *The political ecology of the agrofuels*, London: Routledge, pp. 167–185.

Bailey, I. (2007) Market environmentalism, new environmental policy instruments, and climate policy in the United Kingdom and Germany, *Annals of the Association of American Geographers* 97(3): 530–550.

Bailis, R. and Baka, J. (2011) Constructing sustainable biofuels: Governance of the emerging biofuel economy. *Annals of the Association of American Geographers* 101(4): 827–838.

Baines, J. (2015) Fuel, feed and the corporate restructuring of the food regime, *The Journal of Peasant Studies* 42(2): 295–321.

Balan, V., Chiaramonti, D. and Kumar, S. (2013) Review of US and EU initiatives toward development, demonstration, and commercialization of lignocellulosic biofuels, *Biofuels, Bioproducts and Biorefining* 7(6): 732–759.

Bello, W. (2009) *The food wars*, London: Verso Books.

Birch, K. (2006) The neoliberal underpinnings of the bioeconomy: The ideological discourses and practices of economic competitiveness, *Genomics, Society and Policy* 2(3): 1–15.

Birch, K. (2016a) Emergent policy imaginaries and fragmented policy frameworks in the Canadian bio-economy, *Sustainability* 8(10): 1–16.

Birch, K. (2016b) Materiality and sustainability transitions: Integrating climate change into transport infrastructure in Ontario, Canada, *Prometheus: Critical Studies in Innovation* 34(3–4): 191–206.

Birch, K. (2017) *A research agenda for neoliberalism*, Cheltenham: Edward Elgar.

Birch, K. and Calvert, K. (2015) Rethinking 'drop-in' biofuels: On the political materialities of bioenergy, *Science and Technology Studies* 28(1): 52–72.

Birch, K., Levidow, L. and Papaioannou, T. (2010) Sustainable capital? The neo-liberalization of nature and knowledge in the European "knowledge-based bio-economy", *Sustainability* 2: 2898–2918.

Birch, K., Levidow, L. and Papaioannou, T. (2014) Self-fulfilling prophecies of the European knowledge-based bio-economy: The discursive shaping of institutional and policy frameworks in the bio-pharmaceuticals sector, *Journal of the Knowledge Economy* 5: 1–18.

Birner, R. (2018) Bioeconomy concepts, in I. Lewandowski (eds) *Bioeconomy*, Cham: Springer, pp. 17–38.

Bonaccorso, M. (2015) An interview with the "father" of EU bioeconomy, *ilbioeconomista.com*, retrieved from: https://ilbioeconomista.com/2015/11/19/chris-patermann-talks-to-il-bioeconomista-an-interview-with-the-father-of-eu-bioeconomy/.

Bonneuil, C. and Fressoz, J.-B. (2017) *The shock of the anthropocene*, London: Verso.

Boykoff, M. and Randalls, S. (2009) Theorizing the carbon economy: Introduction to the special issue, *Environment and Planning A* 41(10): 2299–2304.

Bridge, G. (2009) Material worlds: Natural resources, resource geography and the material economy, *Geography Compass* 3(3), 1217–1244.

Bridge, G. (2010) Geographies of peak oil: The other carbon problem, *Geoforum* 41: 523–530.

Bugge, M., Hansen, T. and Klitkou, A. (2016) What is the bioeconomy? A review of the literature, *Sustainability* 8: 1–22.

Calvert, K. (2016) From 'energy geography' to 'energy geographies': Perspectives on a fertile academic borderland, *Progress in Human Geography* 40(1): 105–125.

Calvert, K., Birch, K. and Mabee, W. (2017) New perspectives on an old energy resource: Biomass and emerging bio-economies, in S. Bouzarovski, M. Pasqualetti and V. Castan Broto (eds) *The Routledge research companion to energy geographies*, London: Routledge, pp. 47–60.

CEC. (2005) *New perspectives on the knowledge-based bio-economy: Conference report*, Brussels: DG-Research, Commission of the European Communities.

CEC. (2012) *Innovating for sustainable growth: A bioeconomy for europe* [COM(2012) 60 Final], Brussels: Commission of the European Communities.

Charles, M., Ryan, R., Ryan, N. and Oloruntoba, R. (2007) Public policy and biofuels: The way forward? *Energy Policy* 35(11): 5737–5746.

ClientEarth. (2012) *Carbon impacts of bioenergy under European and international rules*. Brussels: ClientEarth.

Dahmann, K., Fowler, L. and Smith, P. (2016) United States law and policy and the biofuel industry, in Y. Le Bouthillier, A. Cowie, P. Martin, and H. McLeod-Kilmurray (eds) *The law and policy of biofuels*, Cheltenham: Edward Elgar, pp. 102–140.

de Besi, M. and McCormick, K. (2015) Towards a bioeconomy in Europe: National, regional and industrial strategies, *Sustainability* 7: 10461–10478.

El-Chickakli, B., von Braun, J., Lang, C., Barben, D. and Philp, J. (2016) Five cornerstones of a global bioeconomy, *Nature* 535: 221–223.

ePure. (2017) European renewable ethanol – key figures 2016, *epure.org*, retrieved from: http://epure.org/media/1610/2016-industry-statistics.pdf.

Fargione, J., Hill, J., Tilman, D., Polasky, S. and Hawthorne, P. (2008) Land clearing and the biofuel carbon debt, *Science* 319: 1235–1238.

Felli, R. (2014) On climate rent, *Historical Materialism* 22(3–4): 251–280.

Foucault, M. (2008) *The birth of biopolitics*, New York: Picador.

Frow, E., Ingram, D., Powell, W., Steer, D., Vogel, J. and Yearley, S. (2009) The politics of plants, *Food Security* 1(1): 17–23.

Hausknost, D., Schriefl, E., Lauk, C. and Kalt, G. (2017) A transition to which bioeconomy? An exploration of diverging techno-political choices, *Sustainability* 9(4): 1–22.

HLPE. (2013) *Biofuels and food security: A report by the high level panel of experts on food security and nutrition*, Rome: Committee on World Food Security.

German Bioeconomy Council. (2015a) *Bioeconomy policy: Synopsis and analysis of strategies in the G7*, Berlin: Office of the German Bioeconomy Council.

German Bioeconomy Council. (2015b) *Bioeconomy policy: Synopsis of national strategies around the world*, Berlin: Office of the German Bioeconomy Council.

Gillon, S. (2010) Fields of dreams: Negotiating an ethanol agenda in the Midwest United States, *The Journal of Peasant Studies* 37: 723–748.

Goldstein, J. (2018) *Planetary improvement*, Cambridge, MA: MIT Press.

Goldstein, J. and Tyfield, D. (2017) Green Keynesianism: Bringing the entrepreneurial state back in(to question)?, *Science as Culture*, doi.org/10.1080/09505431.2017.1346598.

Goven, J. and Pavone, V. (2015) The bioeconomy as political project: A Polanyian analysis, *Science, Technology, and Human Values* 40(3): 302–337.

Hilgartner, S. (2007) Making the bioeconomy measurable: Politics of an emerging anticipatory machinery, *BioSocieties* 2: 382–386.

Hilgartner, S. (2015) Capturing the imaginary: Vanguards, visions and the synthetic biology revolution, in S. Hilgartner, C. Miller and R. Hagendijk (eds) *Science and democracy*, London: Routledge, pp. 33–55.

Hill, J., Nelson, E., Tilman, D., Polasky, S. and Tiffany, D. (2006) Environmental, economic, and energetic costs and benefits of biodiesel and ethanol biofuels, *PNAS* 103(30): 11206–11210.

HM Treasury. (2006) *Stern review on the economics of climate change*, London: HM Treasury, retrieved from: http://webarchive.nationalarchives.gov.uk/+/http://www.hm-treasury.gov.uk/sternreview_index.htm.

IEA. (2011) *Technology roadmap: Biofuels for transport*, Paris: International Energy Agency.

IEA. (2017) *Technology roadmap: Delivering sustainable bioenergy*, Paris: International Energy Agency.

Irwin, S. and Good, D. (2017) The EPA's renewable fuel standard rulemaking for 2018 – Still a push, *farmdoc daily* 7: 125, retrieved from: http://farmdocdaily. illinois.edu/2017/07/epa-renewable-fuel-standard-rulemaking-for-2018. html.

Jackson, T. (2009) *Prosperity without growth*, London: Earthscan.

Kircher, M. (2012) The transition to a bio-economy: National perspectives, *Biofuels, Bioproducts & Biorefining* 6(3): 240–245.

Kitchen, L. and Marsden, T. (2011) Constructing sustainable communities: A theoretical exploration of the bio-economy and eco-economy paradigms, *Local Environment* 16: 753–769.

Le Bouthillier, Y., Cowie, A., Martin, P. and McLeod-Kilmurray, H. (eds) (2016) *The law and policy of biofuels*, Cheltenham: Edward Elgar.

Le Quéré, C. et al. (2017) Global Carbon Budget 2017, *Earth Systems Science Data Discussions*, retrieved at: https://doi.org/10.5194/essdd-2017-123.

Levidow, L., Birch, K. and Papaioannou, T. (2012) EU agri-innovation policy: Two contending visions of the knowledge-based bio-economy, *Critical Policy Studies* 6: 40–66.

Levidow, L., Birch, K. and Papaioannou, T. (2013) Divergent paradigms of European agro-food innovation: The knowledge-based bio-economy (KBBE) as an R&D agenda, *Science, Technology, and Human Values* 38: 94–125.

Liverman, D. (2009) Conventions of climate change: Constructions of danger and the dispossession of the atmosphere, *Journal of Historical Geography*, 35(2): 279–296.

Lohmann, L. (2010) Neoliberalism and the calculable world: The rise of carbon trading, in K. Birch and V. Mykhnenko (eds) *The rise and fall of neoliberalism*, London: Zed Books, pp. 77–93.

Lohmann, L. (2016) Neoliberalism's climate, in S. Springer, K. Birch and J. MacLeavy (eds.) *The handbook of neoliberalism*, London, Routledge, pp. 480–492.

Londo, M. and Deurwaarder, E. (2007) Developments in EU biofuels policy related to sustainability issues: Overview and outlook, *Biofuels, Bioproducts and Biorefining* 1(4): 292–302.

Mathews, J.A. (2009) From the petroeconomy to the bioeconomy: Integrating bioenergy production with agricultural demands, *Biofuels, Bioproducts & Biorefining* 3: 613–632.

McCormick, K. (2014) The bioeconomy and beyond: Visions and strategies, *Biofuels* 5: 191–193.

McCormick, K. and Kautto, N. (2013) The bioeconomy in Europe: An overview, *Sustainability* 5: 2589–2608.

McKibben, B. (2012) Global warming's terrifying new math, *Rolling Stone* 19 July, retrieved from: https://www.rollingstone.com/politics/news/global-warmings-terrifying-new-math-20120719.

McKibben, B. (2016) Recalculating the climate math, *New Republic* 22 September, retrieved from: https://newrepublic.com/article/136987/recalculating-climate-math.

McMichael, P. (2009) The agrofuels project at large, *Critical Sociology* 35(6): 825–839.

Mitchell, D. (2008) *A note on rising food prices*, Washington, DC: World Bank.

Mitchell, T. (2011) *Carbon democracy*, London: Verso.

Moore, J.W. (2015) *Capitalism in the web of life*, London: Verso.

Mukhtarov, F., Gerlak, A. and Pierce, R. (2017) Away from fossil-fuels and toward a bioeconomy: Knowledge versatility for public policy? *Environment and Planning C* 35(6): 1010–1028.

OECD. (2004) *Biotechnology for sustainable growth and development*, Paris: Organisation for Economic Co-operation and Development.

OECD. (2005) *The bioeconomy to 2030: Designing a policy agenda*, Paris: Organisation for Economic Co-operation and Development.

OECD. (2006) *The bioeconomy to 2030: Designing a policy agenda*, Paris: Organisation for Economic Co-operation and Development.

OECD. (2009) *The bioeconomy to 2030: Designing a policy agenda – Main findings and policy conclusions*, Paris: Organisation for Economic Co-operation and Development.

Olsen, B.E. and Rønne, A. (2016) The EU legal regime for biofuels, in Y. Le Bouthillier, A. Cowie, P. Martin, and H. McLeod-Kilmurray (eds) *The law and policy of biofuels*, Cheltenham: Edward Elgar, pp. 164–190.

Oreskes, N. and Conway, E. (2010) *Merchants of doubt*, New York: Bloomsbury.

Ostergard, H., Markussen, M. and Meeusen, M. (2010) Challenges for sustainable development, in H. Langeveld, J. Sanders and M. Meeusen (eds) *The bio-based economy: Biofuels, materials and chemicals in the post-oil era*, Bristol: Earthscan, pp. 33–49.

Peck, J. (2010) *Constructions of neoliberal reason*, Oxford: Oxford University Press.

Pfau, S.F., Hagens, J.E., Dankbaar, B. and Smits, A.J.M. (2014) Visions of sustainability in bioeconomy research, *Sustainability* 6: 1222–1249.

Pimentel, D. (2009) Biofuel food disasters and cellulosic ethanol problems, *Bulletin of Science, Technology and Society* 29(3): 205–212.

Pimentel, D. and Patzek, T. (2006) Green plants, fossil fuels, and now biofuels, *Bioscience* 56(11): 875.

Ponte, S. (2014) The evolutionary dynamics of biofuel value chains: From unipolar and government-driven to multipolar governance, *Environment and Planning A* 46(2): 353–372.

Ponte, S. and Birch, K. (2014) Introduction: The imaginaries and governance of 'biofueled futures', *Environment and Planning A* 46(2): 271–279.

Power, M. (2016) Lessons from US biofuels policy: The renewable fuel Standard's rocky ride, in Y. Le Bouthillier, A. Cowie, P. Martin, and H. McLeod-Kilmurray (eds) *The law and policy of biofuels*, Cheltenham: Edward Elgar, pp. 141–163.

Richardson, B. (2012) From a fossil-fuel to a biobased economy: The politics of industrial biotechnology, *Environment and Planning C* 30: 282–296.

Rico, J.A.P. (2008) *Programa de biocombustíveis no Brasil e na Colômbia: Uma análise da implantação, resultados e perspectivas* (in Portuguese), Universidade de São Paulo: PhD Thesis, pp. 81–82.

Robbins, P., Hintz, J. and Moore, S. (2010) *Environment and society*, Oxford: Wiley-Blackwell.

Scarlat, N., Dallemand, J.-F., Monforti-Ferraro, F. and Nita, V. (2015) The role of biomass and bioenergy in a future bioeconomy: Policies and facts, *Environmental Development* 15: 3–34.

Schmid, O., Padel, S. and Levidow, L. (2012) The bio-economy concept and knowledge base in a public goods and farmer perspective, *Bio-based and Applied Economics* 1: 47–63.

Schnepf, R. and Yacobucci, B. (2013) *Renewable Fuel Standard (RFS): Overview and issues*, Congressional Research Service, 7-5700, retrieved from: https://www.ifdaonline.org/IFDA/media/IFDA/GR/CRS-RFS-Overview-Issues.pdf.

Searchinger, T., Heimlich, R., Houghton, R.A., Dong, F., Elobeid, A., Fabiosa, J., Tokgoz, S., Hayes, D. and Yu, T.-H. (2008) Use of US croplands for biofuels increases greenhouse gases through emissions from land-use change, *Science* 319: 1238–1240.

Shiva, V. (2008) *Soil not oil*, Cambridge, MA: South End Press.

Shortall, O., Raman, S. and Millar, K. (2015) Are plants the new oil? Responsible innovation, biorefining and multipurpose agriculture, *Energy Policy* 86: 360–368.

Sims, R., Mabee, W., Saddler, J. and Taylor, M. (2010) An overview of second generation biofuel technologies, *Bioresource Technology* 101: 1570–1580.

Smith J (2010) *Biofuels and globalization of risk.* London: Zed Books.

Sorda, Gi., Banse, M. and Kemfert, C. (2010) An overview of biofuel policies across the world, *Energy Policy* 38 (11): 6977–6988.

Staffas, L., Gustavsson, M. and McCormick, K. (2013) Strategies and policies for the bioeconomy and bio-based economy: An analysis of official national approaches, *Sustainability* 5: 2751–2769.

Stephen, J., Mabee, W. and Saddler, J. (2011) Will second-generation ethanol be able to compete with first-generation ethanol? Opportunities for cost reduction, *Biofuels, Bioproducts and Biorefining* 6(2): 159–176.

Tickell, O. (2009) *Kyoto2*, London: Zed Books.

Trottiers. (2013) *Low-carbon energy futures: A review of national scenarios*, Trottiers Energy Future Project.

Tyfield, D. (2014) 'King coal is dead! Long live the king!': The paradoxes of coal's resurgence in the emergence of global low-carbon societies, *Theory, Culture & Society* 31(5): 59–81.

Tyner, W. (2010a) Cellulosic biofuels market uncertainties and government policy, *Biofuels* 1(3): 389–391.

Tyner, W. (2010b) Comparisons of the US and EU approaches to stimulating biofuels, *Biofuels* 1(1): 19–21.

Unruh, G. (2000) Understanding carbon lockin, *Energy Policy* 28(12): 817–830.

van Meijl, H., Smeets, E. and Zilberman, D. (2015) Bioenergy economics and policies, in R. Victoria, C. Roly, and L. Verdade (eds) *Bioenergy and sustainability*, Paris: SCOPE, pp. 682–708.

White House. (2012) *National bioeconomy blueprint*, Washington, DC: The White House.

Worldwatch Institute. (2007) *Biofuels for transport*, London: Earthscan.

Wood, T. and Mabee, W. (2016) *ACW: Baseline report – Energy*, Working Paper #ACW-04, York University.

Yue, D., You, F. and Snyder, S. (2014) Biomass-to-bioenergy and biofuel supply chain optimization: Overview, key issues and challenges, *Computers and Chemical Engineering* 66: 36–56.

Zimmerer, K. (2011) New geographies of energy: Introduction to the special issue, *Annals of the Association of American Geographers* 101(4): 705–711.

Zwart, H., Krabbenborg, L. and Zwier, J. (2015) Is dandelion rubber more natural? Naturalness, biotechnology and the transition towards a bio-based society *Journal of Agricultural and Environmental Ethics* 28: 313–334.

CHAPTER 4

Bio-Economy Policy Visions

INTRODUCTION

Today, examples of the bio-economy are increasingly common all around us, including the use of biomass to create liquid biofuels or biochemicals to the idea of using plants to create biodegradable plastics (e.g. Coca-Cola's PlantBottle®). The very notion of a 'bio' economy has been gaining ground, at least in policy circles, since the mid-2000s when it first appeared on the international policy scene with the support of the Organisation for Economic Co-operation and Development and European Commission (OECD 2005, 2006; CEC 2005). Since then, the bio-economy has become an increasingly important policy vision and agenda around the world; for example, policy-makers in countries like the USA, Germany, and Japan have all developed dedicated bio-economy policy strategies (German Bioeconomy Council 2015a, b).

The bio-economy is generally used in these policy circles to refer to a range of industrial sectors producing biological products or resources, from healthcare through forestry to agriculture. This has led to the conceptual conflation of the terms 'bio-economy' and 'bio-based economy' (McCormick and Kautto 2013). Increasingly, however, it is being used to refer to a possible societal transition (see Geels 2002) from a fossil fuel-based economy to a biological-based economy (e.g. Matthews 2009). More specifically, this transition is premised on the substitution of fossil fuel-based energy, plastics, materials, and chemicals with bio-based ones. Here, the bio-economy is often seen as a win-win-win solution to a series

© The Author(s) 2019   79
K. Birch, *Neoliberal Bio-Economies?*,
https://doi.org/10.1007/978-3-319-91424-4_4

of overlapping societal challenges from climate change to energy security to rural economic development (Frow et al. 2009; Schmid et al. 2012). Part of the (policy) attraction of the bio-economy seems to be its compatibility with existing social institutions and infrastructures—which I come back to in Chap. 5—meaning that it would not require significant changes to social life; for example, ethanol and biodiesel slot into existing petroleum value chains with little need for change. As such, it's framed as a societal 'transition' but not a societal 'transformation'.

In this chapter, I conceptualize the bio-economy as a *policy framework* involving a vision or imaginary of the future and a particular set of proposed policies and institutional changes for achieving that vision (Birch et al. 2010, 2014; Levidow et al. 2012, 2013; Birch 2016). Initially, the bio-economy concept was used to refer to the "the recent surge in the scientific knowledge and technical competences" (OECD 2006: 3), or the idea that "knowledge has become an extremely valuable economic resource" (CEC 2005: 1). The particularities of these new (bio-)knowledges, however, meant that policy-makers faced significant difficulties in managing the impacts and outcomes of scientific research and innovation, especially in their desire to capture future market opportunities. For example, the OECD (2006: 5) argued that:

> The challenge facing policy makers – whether in government, in private industry or elsewhere – is how to make choices that allow the opportunities offered through biotechnology, genetics, genomics and the biosciences more generally to be delivered. This can be problematic, since decisions taken today can influence whether unforeseen or unconfirmed future opportunities might be realizable. A degree of foresight or vision is therefore necessary so that, to the extent possible, short-term decisions can be taken without negative impacts on longer term opportunities.

In this sense, and as Hilgartner (2007) notes, the OECD's original bio-economy concept represented an 'anticipatory machine' or a 'future-making project', rather than a simple description or outline of a new sector, new product market, or new resource base. According to several other scholars (e.g. Birch et al. 2010, 2014; Levidow et al. 2012, 2013; Birch 2016), the bio-economy concept has been used more broadly to discursively frame *current* policy frameworks and institutions as potentially or actually problematic to the realization of the future market opportunities offered by the biological sciences, necessitating policy and institutional change.

I examine Canada's emerging bio-economy policy vision(s) and their implications for Canadian bio-economy policy frameworks in this chapter. I do so in order to consider what kind of policy visions are evident in the Canadian context and how these definitions inform policy frameworks. My conclusion is that current Canadian bio-economy policy frameworks are fragmented as a result of the emergent and contested nature of Canadian bio-economy policy visions, meaning that future policy and institutional changes deemed necessary to promote the bio-economy are currently stymied. The policy implications of this conclusion are that Canada will need a single, coherent bio-economy vision which a majority of policy stakeholders can buy into before it can develop a policy strategy to promote and support the bio-economy.

I start the chapter with a discussion of my conceptual approach linking policy visions with policy frameworks, which drives policy and institutional change. In the empirical analysis, I focus on the emergence of diverse, often competing visions of the bio-economy in the Canadian context and the fragmentation of policy frameworks that results from this diversity. I then conclude.

IMAGINED FUTURES: POLICY VISIONS AND POLICY FRAMEWORKS

Markets and natures are not givens; they are made and to be made they have to be envisioned and imagined. As the discussion of neoliberal natures in Chap. 2 illustrates, however, the role of visions and imaginaries in the construction of markets and natures is often obscured in the analyses of the deployment of market-based mechanisms in environmental policy-making or sustainable business practices. Visions, imaginaries, discourses, and narratives—especially of the market-centred kind—are frequently conceptualized as drivers of market-based decision-making (e.g. neoliberal ideology), rather than something that needs to be explained (cf. Birch et al. 2010). According to the environmental economic geography (EEG) literature, however, the construction of nature-economy relations is as much a discursive process as it is material (e.g. Bridge 2008, 2009; Soyez and Schulz 2008). In particular, visions and imaginaries are not simply descriptive narratives, they're also performative and generative in that they direct policy attention, enrol material and social support, and create awareness of different understandings of the world and their manifestations (Hilgartner 2007; Birch et al. 2010, 2014; Ponte and Birch 2014; Bauer

2018). As I noted already, for example, future visions of what sorts of society we want to transition to in the future come to define how we understand nature and natural resources—as waste, as unused residue, as a circular input, and so on. Looking at these cultural attitudes, beliefs, and suchlike is, therefore, central to any understanding of how the bio-economy is organized and governed.

As mentioned in Chap. 3, an increasing number of national governments have or are developing strategies to support and promote the bio-economy (e.g. CEC 2005, 2012; OECD 2006, 2009; White House 2012; German Bioeconomy Council 2015a, b). For example, the European Union's (EU) *Horizon 2020* strategy—a replacement for the Lisbon Agenda (Birch and Mykhnenko 2014)—includes the bio-economy as a way to achieve "a more innovative, resource efficient and competitive society" (CEC 2012: 2). Several scholars have analysed some of these policy strategies, including Staffas et al. (2013), who reviewed various national strategies, and de Besi and McCormick (2015), who reviewed a number of national, regional, and industry strategies.

These policy strategies are constituted by both visions of the future—often involving competing social, political, economic, and scientific priorities—*and* frameworks for achieving those visions, creating a performative driver towards the *imagined* future (Hilgartner 2007). As such, policy strategies enrol support, redirect resources, shape regulations, and create markets in a diverse range of policy areas including research, innovation, industry, energy, transport, and agriculture. At present, dominant policy strategies tend to imagine the bio-economy as a sustainable system in which biological materials (e.g. biomass) can relatively easily replace fossil fuels as the underlying resource base for our societies and economies (Richardson 2012; McCormick 2014). As such, the bio-economy is framed as an important and cost effective socio-technical transition pathway leading us to a sustainable future (Geels 2002). Key elements in this transition pathway are the development of new forms of energy (e.g. liquid biofuels), new forms of intermediate inputs (e.g. biochemicals), and new forms of products (e.g. bioplastics) (Schmid et al. 2012; Ponte and Birch 2014; Birch and Calvert 2015; Coenen et al. 2015; Bauer 2018).

Theoretically, I build on my previous work, including Birch et al. (2010, 2014), Levidow et al. (2012, 2013), Ponte and Birch (2014), and Birch (2016), in order to conceptualize the bio-economy as a particular policy vision and framework (also Hilgartner 2007). It represents a techno-economic imaginary of the future that is co-produced with certain policies, institutions, and infrastructures that are framed as desirable *and* possible, while others are framed as undesirable or problematic. It's important, from this perspective, to examine and analyse the bio-economy as it is constituted by imaginaries and policy frameworks, reflecting an overall policy strategy designed to (re)configure a particular techno-economic regime or system.

Starting with *imaginaries*, these have a long conceptual history stretching back to Cornelius Castoriadis, Benedict Anderson, Charles Taylor, and George Marcus. More recently, Sheila Jasanoff (2004) has analysed the co-production of socio-technical imaginaries and national science policy. Others, like Bob Jessop (2005), have analysed economic imaginaries (e.g. 'knowledge-based economy') as discourses that underpin particular regimes of accumulation (e.g. post-Fordism). Generally, this research highlights the role of discourses, narratives, and visions in the (re)configuration of socio-economic phenomena like scientific research or economic activity (also Felt et al. 2007). For example, imaginaries provide a conceptual tool to analyse how social actors understand policies, how policy frameworks emerge, and the social actors involved, or excluded, from the development of future visions of society.

Imaginaries and *policy frameworks* are co-constituted, enrolling a range of stakeholders in the pursuit of particular policy strategies (Ponte and Birch 2014). Policy frameworks can be conceptualized as the policy priorities, analysis, funding, schemes, initiatives and directives, and implementation modes that cut across policy-making, ensuring that policies are compatible and complementary (Birch et al. 2010). Analysing policy frameworks helps to understand (1) how policy imaginaries are aligned with a particular and often prevailing configuration of socio-economic institutions and infrastructures (Birch and Calvert 2015) and (2) how this prevailing configuration might be opened up and alternatives adopted (Birch et al. 2010; Schmid et al. 2012; Levidow et al. 2013). On the one hand, scholars have noted how dominant bio-economy visions and policy frameworks tend to be so broadly conceived that they represent 'something for everyone'—for example, sustainability, energy security, rural development, and so on (Frow et al. 2009; Richardson 2012). On the other hand, a number of scholars have been critical of how dominant policy visions of the bio-economy are centred on finding technoscientific solutions to societal problems, rather than rethinking the social, political, economic, and ecological systems that bio-economy policies are supposed to transform (Kitchen and Marsden 2011).

In light of this theoretical discussion, I have two analytical objectives in this chapter. First, while there is now a growing literature discussing the bio-economy (e.g. Staffas et al. 2013; Pfau et al. 2014; Bugge et al. 2016; El-Chichakli et al. 2016; Bauer 2018), much of it only focuses on certain jurisdictions or jurisdictional examples—especially the European Union (EU) and its member states—and certain policy actors, especially government actors. There is, therefore, a need to examine a broader array of bio-economy strategies and their constitutive elements and actors in depth. By analysing the Canadian bio-economy, and its range of government, industry, and civil society stakeholders, I aim to provide a useful comparative perspective to existing work, which focuses mostly on Europe (e.g. Birch et al. 2010, 2014; Levidow et al. 2012, 2013).

Second, in the extant literature on imaginaries (e.g. Jasanoff 2004; Jessop 2005), the examples of policy visions, narratives, and discourses (or imaginaries for short) analysed are generally dominant or stabilized imaginaries; while they may have once been emergent, the analysis tends towards explaining stable and dominant imaginaries, rather than analysing the diversity and range of potential or emergent visions before a dominant imaginary stabilizes. An example is Jasanoff and Kim's (2009) work on nuclear imaginaries in South Korea and the USA; they describe the dominant imaginaries of each nation-state, rather than the potential imaginaries that could have stabilized. Obviously, this analytical focus results from the practicalities of the research process; for example, Jasanoff and Kim had to take a retrospective approach in their analysis, since they were examining historical events. In contrast, I analyse emergent imaginaries in this chapter, each of which could, potentially, become dominant or stabilized in a particular policy framework. It's possible that a single (Canadian) imaginary will emerge from a range of different, diverse, and even competing visions, but it's unclear what this imaginary will entail or when it will stabilize.

Emergent Imaginaries in the Canadian Bio-Economy

Imaginaries are often geographically specific, in that different countries are characterized by different visions of the future. For example, when it comes to the bio-economy in Canada, future visions have been driven, generally speaking, by a radical shift in the Canadian economy away from the traditional (self-)conception of Canada as 'hewers of wood' to one based on scientific research and high-tech development (e.g. CRFA 2014). It's not my intention to analyse these geographical imaginaries in this chapter however (see Birch 2012; Ponte and Birch 2014). Instead, my aim in this section is to analyse the different—sometimes overlapping and sometimes competing—definitions of the bio-economy in Canadian policy context. I outline four definitions of the bio-economy used by various policy actors and analyse the origins of these definitions and their relationships to particular policy visions. The four definitions are: (1) *product*-based; (2) *substitution*; (3) *renewable*-versus-*sustainable*; and (4) *societal transitions*.

Competing Definitions of the Bio-Economy 1: Bio-based Products

One of the dominant definitions used by Canadian policy actors is the idea that the bio-economy represents the development and manufacture of new bio-based products and energy; examples range from the low-tech end with wood-based disposable cutlery, through bulk energy resources like ethanol or cellulosic biofuels, to high-end chemicals like succinic acid. The key defining characteristic of the bio-economy here is that these products, energy, and so on must be manufactured using biological inputs (e.g. agricultural waste, timber, etc.). For example, several informants described the bio-economy in the following ways:

- Consultancy (#93): "It would be the manufacturing use of products that are made from biological organisms. So things that are grown, can live, and be harvested and sustainably used again through the life-cycle of an organism."
- Trade Association (#99): "From our vantage point we view the bio-economy as any product or thing, if you will, that can be produced from crops that isn't traditionally food."
- Consultancy (#100): "Making stuff from carbohydrates ... So basically, I'm saying that manufacturing products that come in whole or in part from biomass."
- Federal Agency (#106): "For me the bio-economy is the development of technologies and products that are derived from things that use as an input naturally occurring materials ... But something that comes from an organic nature."
- Provincial Ministry (#110): "I think bio-economy composes of anything that includes the use of either whole of significant part of any biological renewable materials or any organic materials from plants and animals which could be used to make a product for commercial or industrial use."

One informant, a civil society actor in this case, went so far as to describe this *product* definition as the 'standard definition', suggesting that it is, perhaps, emerging as a dominant policy imaginary.

Civil Society Organization (#95): "Well we use sort of a standard definition where basically all materials, energy, chemicals and consumer products are built out of or based upon renewable biological resources, as opposed to non-renewable fossil fuels."

As a result of this focus on bio-based products, the broader policy vision tends to emphasize the *economic* aspects of the bio-economy, rather than the social, political or sustainable aspects. As such, it reflects other bio-economy policy visions elsewhere, especially those developed by the OECD, EU, and USA (OECD 2009; CEC 2012; White House 2012). As McCormick (2014) notes, these visions emphasize the following:

- OECD (2009): development of "sustainable, eco-efficient and competitive products."
- CEC (2012): development of new products and integration of biological resources into the economy.
- USA (2012): use of (biological) science and innovation to create economic growth.

Such definitions mean that only certain policy strategies are deemed desirable or possible. For example, the emphasis on markets frames the use of certain measures (e.g. government incentives) to promote the bio-economy in certain ways (e.g. to create markets). Moreover, it frames the bio-economy normatively as driven by particular interests (e.g. consumers) at the expense of other concerns (e.g. ecological protection) (Kitchen and Marsden 2011).

Generally, how the future is defined and framed ends up shaping potential policy strategies and policy-making, especially when it comes to innovation and market development policies, and the various institutions and infrastructures on which these policies depend (see Chap. 6). As noted elsewhere in the literature, this then legitimates the shaping of policies and institutions to suit the future vision (Birch et al. 2014), and a topic I return to in Chap. 5. It's notable that this 'product' vision, while it might promote policy and institutional change, is underpinned by an expectation that any policy strategy would integrate bio-based products, energy, and so on into the prevailing political economy, rather than challenging it with alternative political-economic paradigms (e.g. circular, degrowth, steady state, etc.). For example, several informants outlined this compatibility with prevailing market-based expectations in the following ways:

- Trade Association (#97): "So as far as the bio-economy goes, I mean, I think from our perspective we see it as an outlet or opportunity for-, to take demand for our product, essentially."
- Trade Association (#107): "I think the bio-economy is leveraging the renewable resources that we're blessed with in Canada to deliver sustainable products to global markets."
- Trade Association (#109): "...about creating an environment where sustainable products and stable fuels can be more easily integrated into the economy of Canada. It's a comprehensive approach to the management of our resources in a way that promote the use of less carbon intensive fuels."

As the quotes above illustrate, the product definition is embedded within a *market-based* or *neoliberal* vision of the bio-economy as a new sector, which extends market-based understandings and their implementation to new *and* old resource areas. It is not, in this framing, a threat to existing political economy or to incumbent firms. This definition is the main way that industry actors frame the bio-economy; notably, it is also a common way that government actors frame it, although civil society actors do not use this definition by and large. In terms of the Canadian context, this definition is likely a cause *and* effect of the fragmented policy framework in Canada (see below), which often ends up with policy initiatives and instruments being directed at narrow industrial sectors or product markets, meaning that it makes sense for both industry and government to define the bio-economy in similarly narrow terms. For example, a Federal Agency (#105) informant noted that "our roadmap is focused on industrial bioproducts stemming from agriculture", rather than from a range of biological resources.

The prevalence of the product-based definition in Canadian policy visions reflects a broader policy tendency to focus on a (neoliberal) techno-fix to solve societal challenges like energy security, climate change, rural development, and so on (Birch et al. 2010). It corresponds to the policy agendas promoted by the OECD, which focus on advances in biotechnological sciences as *the* solution to societal problems (Hilgartner 2007). An example from my research was an informant from Civil Society Association (#91) who defined the bio-economy as "renewable carbon and hydrogen chains to products that benefit society". In considering this product-based representation of the bio-economy in relation to other policy jurisdictions, such definitions are more common in the policy visions of other liberal-market economies like the USA and Australia rather than the more 'bio-based' definitions used in countries in the EU, Germany, Finland, and Sweden (Staffas et al. 2013). This is interesting for the simple reason that Canada's political economy is often characterized as a resource-based one.

Competing Definitions of the Bio-Economy 2: Substitution

Closely aligned, even overlapping, with the product-based definition discussed above is the conception of the bio-economy as an economy defined by the *substitution* of biological products and services for fossil fuel-based ones. Although this definition could be characterized as a product-based vision of the bio-economy, there is one major difference setting it apart. The substitution definition is underpinned, as the name suggests, by the replacement of existing products and services with wholly new products and services, representing a direct challenge, or even threat, to incumbent industry actors and the policy frameworks that support prevailing institutions and infrastructures, if not to markets more generally. Examples of informants describing the bio-economy this way include the following:

- University (#90): "...in a lot of people's minds it's substitution of bio-based material for fossil fuel-based ones."
- Civil Society Association (#94): "...substitution of either a biochemical or a fossil based chemical represents a tremendous gain in terms of footprint."
- Provincial Ministry (#99): "So it's an economic act, we very advanced stage of product using renewable biomass materials to substitute fossil fuel based product and material."
- Federal Agency (#106): "...we are going to be looking for things that can be fully replaced."
- Provincial Ministry (#110): "...substitute fossil based economy with a biobased economy."
- Trade Association (#109): "...we would like to replace those [petroleum based substances] with those that are not based on sequestered carbons."

As these quotes show, this definition is espoused, primarily, by government actors, and in contrast to industry actors. The substitution definition illustrates how different policy visions can be contentious, controversial and contested, meaning that they are less likely to become dominant or stable imaginaries. In the substitution case, while it may share similarities with the product-based definition—in that it also emphasizes the development of products and services—it implicitly and explicitly threatens existing industries (e.g. oil and gas). Consequently, it hasn't become a major policy vision in Canada and is unlikely to do so, as evident in the broader policy discourse. For example, 'substitution' does not appear at all in one of the earliest Canadian bio-economy policy visions, called *Beyond Moose and Mountains: How We Can Build the World's Leading Bio-based Economy* (BIOTECanada 2009). Furthermore, 'fossil' only appears once in that policy vision when the document refers to the idea "that the bio-based economy can – and should – be to the 21st century what the fossil-based economy was to the 20th century" (p. 4). Here, this particular policy vision does not frame the bio-economy as threatening or even competing with incumbent industries, products, or markets, presenting it instead in more general terms.

Another reason that the substitution definition is contentious is that it is, necessarily, equivocal. On the one hand, the development of 'drop-in' bio-based products, services, and energy is the ideal because it means there is no need to replace existing institutions and infrastructures (Birch and Calvert 2015)—see Chap. 5. On the other hand, however, few people think that biomass can replace or supplant fossil fuels across the economy, mainly because the amount of biomass required would be beyond both human and the world's capacity to produce it (Smolker 2008; McCormick and Kautto 2013). As a University (#90) informant argued, "Biomass can't do everything in the economy. You know I don't believe that we are going to have enough of it." Another informant, Civil Society Organization (#92), explicitly argued that the bio-economy should, necessarily, represent an aspect of a future 'hybrid' economy in which fossil fuels also play a significant role. In light of these questions about the potential of bio-based substitution, it's unlikely that the substitution definition can or will become a dominant or stable policy vision for the Canadian bio-economy. However, that doesn't mean that substitution is not important; as a Federal Agency (#105) informant noted, the bio-economy is "a way to sort of improve overall environmental sustainability" through the development of (bio-based) products with less damaging environmental impacts.

Competing Definitions of the Bio-Economy 3: Renewable Versus Sustainable

Across bio-economy policy visions—within and beyond Canada—there is a tendency to conflate 'renewable' with 'sustainable', in the sense that bio-based materials, energy, and products are envisioned as inherently sustainable because they're derived from renewable material. A number of scholars have highlighted this discursive framing in previous research (e.g. Birch et al. 2010; Levidow et al. 2012; Schmid et al. 2012; Pfau et al. 2014). As the previous sub-section illustrates, the bio-economy can then be framed as both sustainable—because it involves replacing fossil fuels with biological matter—and politically feasible, because it involves like-for-like substitutions (e.g. ethanol for petroleum) rather than the wholesale transformation of societies. Examples of this framing include:

- Consultancy (#93): "Renewable is one thing that's sort of inherent in the definition."
- Consultancy (#100): "…you're making products from renewable resources, whether it be trees or crops or whatever, then as long as those trees and crops are managed in a sustainable way, then obviously the bio-economy can be part of a sustainable economy."
- Civil Society Association (#94): "…the bio-economy is an industrial sector that depends on inputs from agriculture, forestry, it means inputs are renewable."

Although this definition is not a dominant policy vision in Canada, it shows how the biophysical qualities of plants can end up being used to define the bio-economy as a 'renewable' policy agenda to sit alongside renewable energy policy more generally. However, this framing is increasingly contested, as illustrated by the debates around the sustainability of first-generation biofuels, especially ethanol, in the USA, Europe, and elsewhere (e.g. Searchinger et al. 2008)—see Chap. 3 for more. For example, research by Gillon (2010, 2014) on the USA and Palmer (2014) on the EU examine contestations around claims about the contribution of biofuels, especially ethanol, to reducing GHG emissions. It's evident, moreover, in the Canadian context that this renewable definition is mainly an 'industry' one, with limited evidence that it is used by government or civil society actors. Even industry actors frame 'renewable' in a nuanced way, as demonstrated by a comment that: "then as long as those trees and crops are managed in a sustainable way, then obviously the bio-economy can be part of a sustainable economy" (Consultancy #100).

As Pfau et al. (2014) highlight, policy research on the bio-economy and sustainability has increasingly emphasized the problems associated with this definition of the bio-economy. Canadian policy actors are similarly sceptical of the notion that the inherent renewable characteristics of plants make them, by definition, sustainable. As one informant noted, this renewable definition is problematic because it means the bio-economy, as a policy, is "not about sustainability anymore, and it's really about biotechnology; in fact, actually, many, but the problem is they imply it's sustainable because it's part of the bio-economy as a bioproduct" (University #96). Amongst other policy actors, there was a strong emphasis, in particular, on the need to integrate sustainability with the bio-economy, rather than see the former as inherent in the latter. Two government informants, for example, stressed that the bio-economy meant "sustainable production of any biomass. So not only for biobased products" (Provincial Ministry #98) and the sustainable use of resources (Provincial Ministry #110). Such definitions were also evident amongst industry and civil society actors as well. One industry informant, for example, said that "I don't think you have the bio-economy without environmental sustainability" (Trade Association #107), implying that they are distinct policy goals. A civil society informant was even clearer on this point, arguing that "when it's done right, when a bio-economy is done in a sustainable fashion, then it is sustainable … biomass is not necessarily sustainable" (Civil Society Organization #95). As these comments illustrate, the idea that the bio-economy is inherently sustainable is highly contested.

Competing Definitions of the Bio-Economy 4: Societal Transitions

A final definition used by Canadian policy actors is the idea that the bio-economy represents a potential socio-technical transition pathway to a low-carbon future (Geels 2002); as with the previous definitions, this definition is not unique to the Canadian context. Although many policy strategies are centred on product-based definitions of the bio-economy (McCormick 2014), according to de Besi and McCormick (2015: 10472), "An important aspect that features across the national and many of the regional [bio-economy] strategies is the need for a transformation in the mind-sets of society, industries and governments." This societal transition definition is evident in the following comments:

- Consultancy (#100): "it's a shift away from the hydrocarbon economy to the carbohydrate economy."
- Civil Society Organization (#95): "for biomass to be sustainable, it has to be renewable within a human lifespan, it has to protect diversity, and it has to value the aesthetics and environmental services that biomass provides while it's living."
- Civil Society Organization (#92): "So you create something that's more of a hybrid system for transition."
- University (#90): "Well you know it's economy, right? So it's spanning not just an individual product or anything, it's all facets of our lives in terms of the way that our society functions and the role that biological resources can play in that."
- University (#103): "trying to shift our economy off the-, well I guess in our case, specifically the petro chemicals, and more onto biologicals that are renewable sustainable biologicals."
- University (#96): "It was one that would move from a – it was like a fossil fuel-based economy to a bio-economy, where biological processes would, we'd use the power of photosynthetic plants and photosynthesis to bring energy and complex carbon molecules into our environment, and in the process not have to … be able to leave some of the fossil fuels in the ground."

As evidenced in these quotes, this definition is used predominantly by civil society stakeholders—it isn't as evident amongst industry informants, if at all. It reflects a more critical understanding of the bio-economy and its sustainability potential, which Pfau et al. (2014) highlight in their work. In their literature review of sustainability and the bio-economy, Pfau et al. argue that one strand of this literature focuses on the negative impacts engendered by the bio-economy, which can include land use change and competition, unintended risks from things like invasive species, disconnection of crops from local ecosystems, and so on. The societal transition definition reflects these sorts of concern in the Canadian context. Specifically, informants emphasized the need to address social, political, and economic issues as part of sustainability concerns; for example, changing social behaviours (e.g. driving less) as well as implementing technoscientific solutions (e.g. advanced biofuels).

Although there are commonalities—such as the desire to make the economy more sustainable—this societal transition definition represents a competing vision of the bio-economy from the other definitions discussed above. As such, it represents an example of the conflict that arises in the development of future visions for our societies, as other analyses of the bio-economy have illustrated elsewhere (e.g. Birch et al. 2010; Levidow et al. 2012, 2013). In the Canadian context, one such conflict arises in the differences between dominant national imaginaries of Canadian society and those of the bio-economy. In particular, many informants, whether industry, government, or civil society actors, stressed the potential—and actual—conflict between incumbent petroleum industrial sectors (e.g. extraction, pipelines, retail, etc.) and the bio-economy. As a number of quotes above show, the societal transitions definition is framed by the need to reduce the extraction and use of fossil fuels as well as develop new institutions (e.g. life cycle analysis standards) and infrastructures (e.g. biomass transportation facilities) to support any societal transition, reflecting a clear difference from the other definitions that try to reduce these social costs through notions of substitution and 'drop-in' biofuels (see Chap. 5). As such, it also reflects a broader interest in concepts like the 'circular economy' and policies supporting renewable energy, recycling, and so on (Mabee 2011).

Fragmented Policy Frameworks in the Canadian Bio-Economy

As I've stressed already, it's important to analyse the policy visions surrounding the bio-economy in order to understand how these imaginaries are co-constitutive of particular policy frameworks. As noted in the theoretical discussion, future visions, narratives, or discourse (i.e. imaginaries) shape and are shaped by policy priorities, funding decisions, implementation processes, and so on (Birch et al. 2010, 2014; Levidow et al. 2012; Ponte and Birch 2014). An important point to remember, in this regard, is that these visions and frameworks have an effect even if the imagined future is not subsequently achieved, since they still engender policy or institutional change (Birch et al. 2014). So even if the purported goals of the policy visions and frameworks aren't realized, they still help to reconfigure policy strategies through institutional changes undertaken to support those strategies. From this perspective, the fact that Canada's bio-economy imaginary is still emergent and contested means that Canada's policy frameworks are likely to be uncoordinated. From my analysis of the interviews, it's evident that Canada's bio-economy policy framework is highly fragmented; moreover, it could be argued that there are multiple policy frameworks. A number of informants argued that there is a need for a single policy vision to create a consistent policy framework in Canada.

Configuring Policy Frameworks

Canadian bio-economy policy frameworks are fragmented as the result of the emergent and contested nature of policy imaginaries, as outlined above. In contrast to the dominant approach that underpins the EU's bio-economy strategy (Birch et al. 2010, 2014; Levidow et al. 2012), for example, Canadian policy visions and frameworks are fragmented by industry and industrial sector. As one informant noted:

> I don't necessarily think that a change in political party is going to make a big change, and I think partly because as a lobby group, the bio-economy players, the agents are just too fragmented. And there's not this over-arching vision about what it can be in terms of who all the players are. (Civil Society Organization #95)

A major reason for this fragmentation is that there is no 'overarching vision' to draw in government, industry, and civil society actors, since each group has their own particular and often competing priorities and objectives. In Canada, moreover, this situation is exacerbated by the fact that industry priorities and objectives often dominate policy-making. For example, a Federal Agency (#105) informant noted that they are concerned mostly with "making sure that we're also aligned with where industry is thinking, we're going in this direction". As another informant from a Federal Agency (#106) pointed out, "they [government] go out to the stakeholders that are representatives of the target industries and say, 'what's missing?'". It's evident that government policy-makers are less concerned with creating an overarching policy vision and framework than with addressing the particular and narrow concerns of different industrial sectors. This contrasts with the situation in the EU and other countries (Staffas et al. 2013; de Besi and McCormick 2015).

Fragmentation is a result of this dominance by industry visions, broadly speaking, in that it leads to a lack of coordination as the different sectoral interests, priorities, and demands come into conflict with one another. This is even the case when different industrial sectors have taken up the term 'bio-economy'; one informant, for example, suggested that "it [bio-economy] has been taken up by a number of companies and groups to actually broaden to any products from biological sources" (University #96). However, although industrial sectors have adopted the term 'bio-economy' in their policy advocacy, this does not mean they use it in the same way. For example, it has been used or adopted by agricultural, food, forestry, biotechnology, bioenergy, biofuels, and bio-product sectors (amongst others). Examples include:

- Forest Products Association of Canada: their *Bio-Pathways Report* (2011) contains no proper definition of 'bio-economy', but refers to "integrating current operations with new add-on processes that create bio-energy, biochemical and bio-materials that add value and jobs" (p. 3).
- Canadian Renewable Fuels Association: their *Evolution and Growth: From Biofuels to Bioeconomy* (2014) report contains, again, no proper definition, but refers to "In addition to environmental benefits, the economic impact of a diverse bioeconomy sector affects the entire value chain, employs a wide array of sciences (life sciences, agronomy, ecology, food science and social sciences), and enables continual development in industrial technologies (biotechnology, nanotechnology and engineering)."
- BIOTECanada: their *Becoming a World Leading Bioeconomy by 2025* (2015) report refers to fact that "The OECD has defined the world bioeconomy as comprising one-third of the total world economy. This includes renewable biomass, and the integration of biotechnology across sectors" (p. 7).

Even when industrial sectors frame the bio-economy in 'product-based' terms, as discussed earlier, this discursive farming doesn't create a consistent or coherent policy vision and framework that all industry sectors can or will adopt. In fact, a product-based policy vision may militate against developing an over-arching policy framework precisely because it is focused on particular and sector-specific products and product strategies (Federal Agency #105). A product-based focus doesn't enable different stakeholders to develop complementary and cross-sectoral priorities or objectives; as one informant noted, "we're incentivizing innovation in a specific and narrow product line for a specific and narrow industry" (Civil Society Organization #95).

A final issue to consider is the extent to which Canadian bio-economy policy frameworks are configured by existing and incumbent industrial sectors, like agriculture, forestry, and energy. While this need not lead to fragmentation, there's a possibility that the orientation of bio-economy policy frameworks towards incumbent and often natural resource sectors (e.g. agriculture, forestry) limits the relevance of the bio-economy for other sectors. One informant claimed:

It seems like each of them sort of had people that were sort of, they were trying to figure out how we're going to use these, the old mills, and of course the bio-economy was a way of doing it. (Federal Ministry #104)

Whether or not this emphasis has configured bio-economy policy visions and frameworks in ways that limit its attractiveness to other sectors is a question worth asking. In Canada, the bio-economy seems to have been seized on as a potential strategy for reinvigorating ailing sectors. For example, one informant argued that "around the latter part of the 1990s the policy platform started to shift and industrial crops were being sought to make things like biofuels to benefit farmers" (Consultancy #93).

Fragmented Policy Frameworks

The fragmentation of bio-economy policy frameworks in Canada can be seen as a consequence of the broader constitutional political structure. In particular, natural resource management falls within provincial jurisdictions, meaning that the national state might only be able to provide limited direction or oversight in promoting an over-arching bio-economy policy vision and framework. As a result, different government agencies and ministries—at the federal and provincial scales—have developed different policy frameworks, if they have one at all. It's worth considering, then, the differences between federal and provincial policy frameworks in order to understand their fragmentation.

At the federal level, a number of informants explicitly stated that Canada doesn't have a bio-economy strategy as such, although policy-makers have been working on one. For example:

- "Well, I mean we don't really have a bio-economy strategy per se, right ... You know you can talk about some bio-economy strategy and stuff, but, you know, then you run into details" (University #90).
- "Canada does not have a national bio-economy strategy ... Yeah we do need to get something like that in place here. And we don't really have it on a provincial basis either" (Civil Society Organization #92).
- "Canadian policy on bio-economy, we really don't have one for agriculture" (Civil Society Association #94).
- "Well, we don't have federally anything that is sort of called a bio-economy policy, at least certainly not one that I'm aware of" (Consultancy #100).

- "Well, certainly you're probably even more familiar than I am of some significant nations have put in place bio-economy policies, Canada has not. That doesn't mean that there aren't really good things in place from a policy perspective that supports bio-economy" (Trade Association #107).
- "We don't have, even in Canada we don't have a federal framework to support bio-economy" (Provincial Ministry #110).

This is not to suggest that there's little support for industry from the federal government; the opposite is the case as an examination of policy initiatives demonstrates. A number of informants provided detailed accounts of various policies, initiatives, and schemes that supported an array of industries—also see Chap. 6. For example, *EcoENERGY for Biofuels and Renewable Power* (est. 2008); *Sustainable Technology Development Canada Fund* and *NextGen Biofuels Fund* (est. 2007); *Renewable Fuel Standard* (est. 2006); and many more (Blair and Mabee 2012). Rather, it's that the bio-economy isn't envisaged as a coherent whole, as one informant explained:

> So, again, the bio-economy for us is the piece that weaves all of our other asks together. The individual asks, we actually get a lot of support for … But an overarching framework just sounds like a lot of work, right? (Trade Association #109)

Another illustration of the fragmentation of Canada's bio-economy strategy is the role played by provincial governments in promoting it. One informant even suggested that "we see this leadership coming-, leadership if you will, coming from the provinces" (Trade Association #97). While provincial policy-makers play a more critical role in supporting and promoting the bio-economy, this merely entrenches the fragmentation of bio-economy policy visions and frameworks as each province develops its own approach based on their different starting positions and plans for the future. For example, Birch and Calvert (2015) provide details of the various mechanisms in place to support bioenergy and biofuels in Ontario, illustrating the role played by provincial governments in supporting new sectors.

In this sense, it's possible to argue that provinces like Ontario have informal and largely implicit bio-economy strategies in place, as one informant claimed. This informant argued that Ontario's policy instruments included a "combination of investment attraction and research that evolved into a provincial policy ... part that we're missing in the provincial policy is the greenhouse gas component" (Civil Society Association #94). A more common outcome of this jurisdictional alignment, however, is the creation of ad hoc policy instruments where "It's almost kind of like by accident you can take something from a province and something from a federal and build something that will help the bio-economy" (Consultancy #100). As such, the onus has been on integrating distributed policy instruments in the pursuit and support of different and distinct industrial sectors, rather than creating an over-arching strategy. Whether or not a national policy vision is actually possible in this context is a question worth asking.

In the Canadian context, it is interesting to note that industry is far more involved in developing a common—if not over-arching—bio-economy policy framework than government, especially in comparison with the EU or other countries (Birch et al. 2010, 2014; Levidow et al. 2012). A number of informants emphasized this point during my research interviews. One informant, for example, claimed that:

> ... there is a sense out there that it needs to be done [to support biofuels], but no, I think honestly, I think the industry, the private sector, is a little more wound up about this right now than the government guys. (University #103)

It's important to note that these interviews were undertaken during Stephen Harper's Conservative administration, which ended in October 2015. The 'neoliberal' orientation of the Conservative government meant that it preferred market-based strategies driven by industry, rather than policy strategies driven by government. As such, there was political uncertainty about the development of bio-economy strategies within government itself, as one industry informant noted: "And the difficulty with that [politics] is it becomes very unpredictable from a policy stand-point. So honestly a lot of our positioning on the bio-economy is defensive" (Trade Association #99). Whether or not a change in government will make a major difference was seen as debatable, however, as another informant noted:

> I don't necessarily think that a change in political party is going to make a big change, and I think partly because as a lobby group, the bio-economy players, the agents are just too fragmented. And there's not this over-arching vision about what it can be in terms of who all the players are. (Civil Society Organization #95)

According to many informants, it's critical for various stakeholders from government, industry, and civil society to collaborate with one another in order to develop a coordinated federal bio-economy strategy. Otherwise, fragmentation is likely to continue. For example, the FPAC (2011), CRFA (2014), and BIOTECanada (2015) reports I cited above all stress the need for such a coordinated policy vision and framework.

Over the last few years, there have been some moves towards such coordination at the Canadian federal level. In particular, a number of informants referred to the creation of a *Bioeconomy Interdepartmental Working Group* (BIWG) in the federal government; created in 2013 or 2014, it's supposed to bring together people from around 15 different federal ministries and agencies. One informant stated that the "overall goal of the group [BIWG] is to develop a Government of Canada strategy for the bio-economy" (Federal Agency #105). As yet, the BIWG hasn't produced any public publications or recommendations, and it seems to be largely focused on fostering collaboration across federal ministries and agencies. At the same time, some federal ministries and agencies have started to create their own working groups as well; for example, *Agriculture and Agri-Food Canada* (AAFC)—itself a member of the BIWG—has established a *Federal-Provincial-Territorial Bioproducts Working Group* and *Industrial Bioproducts Value Chain Round Table*. Despite these initiatives, Canadian policy-makers and stakeholders have as yet to create a coherent national bio-economy policy vision and framework.

CONCLUSION

Over the last few years, many countries, regions, and industries have started to promote and support the bio-economy as an important pathway for societal transition towards a low-carbon and environmentally sustainable future (Geels 2002). The bio-economy has the potential to transform an array of industrial sectors, promising a win-win-win scenario in which energy security, economic growth, and environmental stewardship can go hand-in-hand (Frow et al. 2009). While these policy visions of the bio-economy represent an *imagined* future—in that they don't reflect current technoscientific or political-economic trajectories—they're still powerful drivers of societal change. As Hilgartner (2007) and others suggest, policy visions are powerful because they enrol others in their achievement, they attract resources, they engender popular support, and they side-line possible alternatives. Whether the bio-economy actually makes an impact as imagined in resolving societal challenges like climate change, economic decline, rural development, or energy security could end up as a side issue; it will still have changed policy frameworks and social institutions in important ways.

I sought to explore the bio-economy as a policy strategy co-constituted by policy visions and policy frameworks in this chapter. I built on my previous research, such as Birch et al. (2010, 2014) and Levidow et al. (2012, 2013), and extended this research by analysing the bio-economy in a Canadian context. While others have written about different national bio-economy strategies (e.g. McCormick and Kautto 2013; Staffas et al. 2013; de Besi and McCormick 2015), this chapter—and book more generally—represents the first attempt to dissect the particularities of the Canadian bio-economy. My analysis showed that Canadian policy visions are split between at least four definitions of the bio-economy: product-based, substitution, sustainable-renewable, and societal transition. Moreover, my analysis showed how these different and competing visions of the bio-economy are reflected in the fragmented bio-economy policy frameworks in Canada. As yet, no coherent or coordinated policy framework exists in Canada, but that doesn't mean one cannot be developed.

REFERENCES

Bauer, F. (2018) Narratives of biorefinery innovation for the bioeconomy – Conflict, consensus, or confusion? *Environmental Innovation and Societal Transitions*, doi.org/10.1016/j.eist.2018.01.005.

Birch, K. (2012) Knowledge, place, and power: Geographies of value in the bio-economy, *New Genetics and Society* 31: 183–201.

Birch, K. (2016) *Innovation, regional development and the life sciences: Beyond clusters*, London: Routledge.

Birch, K. and Calvert, K. (2015) Rethinking 'drop-in' biofuels: On the political materialities of bioenergy, *Science and Technology Studies* 28: 52–72.

Birch, K. and Mykhnenko, V. (2014) Lisbonizing vs. financializing Europe? The Lisbon strategy and the (un-)making of the European knowledge-based economy, *Environment and Planning C* 32: 108–128.

Birch, K., Levidow, L. and Papaioannou, T. (2010) Sustainable capital? The neoliberalization of nature and knowledge in the European "knowledge-based bio-economy", *Sustainability* 2: 2898–2918.

Birch, K., Levidow, L. and Papaioannou, T. (2014) Self-fulfilling prophecies of the European knowledge-based bio-economy: The discursive shaping of institutional and policy frameworks in the bio-pharmaceuticals sector, *Journal of the Knowledge Economy* 5: 1–18.

BIOTECanada. (2009) *The Canadian blueprint: Beyond moose & mountains. How we can build the world's leading bio-based economy*, Ottawa: BioteCanada.

BIOTECanada. (2015) *Becoming a world leading bioeconomy by 2025*, Ottawa: BIOTECanada, retrieved from: http://www.biotecanada-ecosystem.com/home-4/.

Blair, J. and Mabee, W. (2012) Policy and legislation to facilitate development of the bioeconomy in Canada, BIOFOR Conference, Thunder Bay, Canada, 14 May.

Bridge, G. (2008) Environmental economic geography: A sympathetic critique, *Geoforum* 39 (1): 76–81.

Bridge, G. (2009) Material worlds: Natural resources, resource geography and the material economy. *Geography Compass* 3 (3): 1217–1244.

Bugge, M., Hansen, T. and Klitkou, A. (2016) What is the bioeconomy? A review of the literature, *Sustainability* 8: 1–22.

CEC. (2005) *New perspectives on the knowledge-based bio-economy: Conference report*, Brussels: DG-Research, Commission of the European Communities.

CEC. (2012) *Innovating for sustainable growth: A bioeconomy for europe* [COM(2012) 60 Final], Brussels: Commission of the European Communities.

Coenen, L., Moodysson, J. and Martin, H. (2015) Path renewal in old industrial regions: Possibilities and limitations for regional innovation policy, *Regional Studies* 49 (5): 850–865.

CRFA. (2014) *Evolution & growth: From biofuels to bioeconomy*, Ottawa: Canadian Renewable Fuels Association.

de Besi, M. and McCormick, K. (2015) Towards a bioeconomy in Europe: National, regional and industrial strategies, *Sustainability* 7: 10461–10478.

El-Chickakli, B., von Braun, J., Lang, C., Barben, D. and Philp, J. (2016) Five cornerstones of a global bioeconomy, *Nature* 535: 221–223.

Felt, U. (rapporteur) (2007) *Taking European knowledge society seriously*, Luxembourg: Office of the Official Publications of the European Communities.

FPAC. (2011) *Bio-pathways report*, Ottawa: Forest Products Association of Canada, retrieved from: http://www.fpac.ca/wp-content/uploads/BIOPATHWAYS-II-web.pdf.

Frow, E., Ingram, D., Powell, W., Steer, D., Vogel, J. and Yearley, S. (2009) The politics of plants, *Food Security* 1 (1): 17–23.

Geels, F. W. (2002) Technological transitions as evolutionary reconfiguration processes: A multi-level perspective and a case-study, *Research Policy* 31 (8–9): 1257–1274.

German Bioeconomy Council. (2015a) *Bioeconomy policy: Synopsis and analysis of strategies in the G7*, Berlin: Office of the German Bioeconomy Council.

German Bioeconomy Council. (2015b) *Bioeconomy policy: Synopsis of national strategies around the world*, Berlin: Office of the German Bioeconomy Council.

Gillon, S. (2010) Fields of dreams: Negotiating an ethanol agenda in the Midwest United States, *The Journal of Peasant Studies* 37: 723–748.

Gillon, S. (2014) Science in carbon economies: Debating what counts in US bio-fuel governance, *Environment and Planning A* 46: 318–336.

Hilgartner, S. (2007) Making the bioeconomy measurable: Politics of an emerging anticipatory machinery, *BioSocieties* 2: 382–386.

Jasanoff, S. (ed) (2004) *States of knowledge: The co-production of science and social order*, London: Routledge.

Jasanoff, S. and Kim, S.-H. (2009) Containing the atom: Sociotechnical imaginaries and nuclear regulation in the U.S. and South Korea, *Minerva* 47: 119–146.

Jessop, B. (2005) Cultural political economy: The knowledge-based economy and the state, in A. Barry and D. Slater (eds) *The technological economy*, London: Routledge, pp. 144–166.

Kitchen, L. and Marsden, T. (2011) Constructing sustainable communities: A theoretical exploration of the bio-economy and eco-economy paradigms, *Local Environment* 16: 753–769.

Levidow, L., Birch, K. and Papaioannou, T. (2013) Divergent paradigms of European agro-food innovation: The knowledge-based bio-economy (KBBE) as an R&D agenda, *Science, Technology, and Human Values* 38: 94–125.

Levidow, L., Birch, K. and Papaioannou, T. (2012) EU agri-innovation policy: Two contending visions of the knowledge-based bio-economy, *Critical Policy Studies* 6: 40–66.

Mabee, W. (2011) *Circular economies and Canada's forest sector*, Work in a Warming World Working Paper #2011–8, York University.

Mathews, J.A. (2009) From the petroeconomy to the bioeconomy: Integrating bioenergy production with agricultural demands, *Biofuels, Bioproducts & Biorefining* 3: 613–632.

McCormick, K. (2014) The bioeconomy and beyond: Visions and strategies, *Biofuels* 5: 191–193.

McCormick, K. and Kautto, N. (2013) The bioeconomy in Europe: An overview, *Sustainability* 5: 2589–2608.

OECD. (2005) *The bioeconomy to 2030: Designing a policy agenda*, Paris: Organisation for Economic Co-operation and Development.

OECD. (2006) *The bioeconomy to 2030: Designing a policy agenda*, Paris: Organisation for Economic Co-operation and Development.

OECD. (2009) *The bioeconomy to 2030: Designing a policy agenda – Main findings and policy conclusions*, Paris: Organisation for Economic Co-operation and Development.

Pfau, S.F., Hagens, J.E., Dankbaar, B. and Smits, A.J.M. (2014) Visions of sustainability in bioeconomy research, *Sustainability* 6: 1222–1249.

Ponte, S. and Birch, K. (2014) Introduction: Imaginaries and governance of bio-fueled futures, *Environment and Planning A* 46: 271–279.

Richardson, B. (2012) From a fossil-fuel to a biobased economy: The politics of industrial biotechnology, *Environment and Planning C* 30: 282–296.

Schmid, O., Padel, S. and Levidow, L.(2012) The bio-economy concept and knowledge base in a public goods and farmer perspective, *Bio-based and Applied Economics* 1: 47–63.

Searchinger, T., Heimlich, R., Houghton, RA., Dong, F., Elobeid, A., Fabiosa, J., Tokgoz, S., Hayes, D. and Yu, TH. (2008) Use of U.S. croplands for biofuels increases greenhouse gases through emissions from land-use change, *Science* 319 (5867): 1238–1240

Smolker, R. (2008) The new bioeconomy and the future of agriculture, *Development* 51: 519–526.

Soyez, D., Schulz, C. (2008) Facets of an emerging Environmental Economic Geography (EEG), *Geoforum* 39 (1): 17–19.

Staffas, L., Gustavsson, M. and McCormick, K. (2013) Strategies and policies for the bioeconomy and bio-based economy: An analysis of official national approaches, *Sustainability* 5: 2751–2769.

The White House. (2012) *National bioeconomy blueprint*, Washington, DC: The White House.

CHAPTER 5

Legitimating Bio-Economies

INTRODUCTION

In a 2018 article for Forbes, the energy expert Robert Rapier outlined the failure of the USA's Renewable Fuel Standard (RFS) to stimulate cellulosic biofuels production. Cellulosic biofuels are integral to RFS2, which was introduced with the 2007 Energy Independence and Security Act (EIS). At the time the legislation was enacted, no commercially viable cellulosic biofuels were being produced. Nevertheless, the cellulosic ethanol mandate—which didn't cover advanced biodiesel production—was set at 100 million gallons in 2010 rising to 16 billion gallons by 2022 (Schnepf and Yacobucci 2013; Dahmann et al. 2016; Power 2016). As Rapier points out, however, no cellulosic biofuels were produced in 2010 or 2011, only 20,000 gallons in 2012, none in 2013, and only 729,000 gallons in 2014, although this jump in production was largely an accounting artefact as biogas was reclassified as cellulosic. By 2017, cellulosic production had reached 10 million gallons, although the original mandate for that year was 5.5 billion gallons—a difference in order of magnitude that makes the actual production quantities largely irrelevant. Despite everything I've written in this book about all the market development policies and policy strategies implemented to transform our societies into low-carbon economies (e.g. subsidies, mandates, etc.), American producers have still

Co-authored with Kirby Calvert

© The Author(s) 2019
K. Birch, *Neoliberal Bio-Economies?*,
https://doi.org/10.1007/978-3-319-91424-4_5

not got anywhere near (economically speaking) the goal of developing an advanced biofuel to replace current fossil fuel use.

Although I come back to this in the next chapter, it's worth noting now that this outcome—the failure so far of this socio-technical transition to a low-carbon future—could be traced to the political-economic materialities of advanced biofuels (see Chap. 2). Attempts to construct a bio-economy come up against a series of biophysical constraints that stymie the efforts and effects of various policies and strategies pursued by different countries, including my empirical focus, Canada. For example, mandates are necessarily bound up with changing infrastructural needs, like the need to increase trucking—versus pipelines—since ethanol evaporates or gels depending on surrounding temperature conditions. The neoliberal natures literature contend that these material limits reflect the recalcitrance of nature, its rebellion against the introduction of market mechanisms in the pursuit of environmental goals (e.g. McCarthy and Prudham 2004; Roff 2008; Bakker 2009). However, as noted in this book already, this perspective ends up framing markets as some sort of aberration of nature—and, in so doing, presenting a 'pristine' nature as an analytical given and something we should be valorizing in our research.

My aim in this book is to examine *how* markets and natures are co-constructed, rather than analysing how markets or natures only constrain, limit, or impinge on one another—even when this involves championing the rebelliousness of natures against neoliberal markets. Consequently, in this chapter, I analyse the legitimation of the bio-economy—and specifically advanced biofuels—as a solution to environmental problems in light of the very biophysical materialities that appear to derail its applicability as a practical societal transition pathway. It means answering some pretty simple questions: how come the bio-economy is an increasingly popular policy strategy (see Chaps. 4 and 6)? How come advanced biofuels are deemed to be a viable technological pathway (see Chap. 6)? My argument is that the bio-economy—or, more accurately, bio-economies in the plural—remains a legitimate policy strategy precisely because it is compatible with incumbent and prevailing political-economic materialities; for example, advanced biofuels are popular because they are designed to be biophysically compatible with existing institutions, infrastructures, and value chains. A key goal, in this case, is the pursuit of so-called 'drop-in' (advanced) biofuels, which provides a legitimating vision, strategy, and policy suite for the bio-economy, or at least one particular version of the bio-economy. More specifically, the bio-economy and its component techno-fixes (e.g. advanced biofuels) are politically popular—as well as strongly supported—because they do not entail a wholesale social, technical, *or* material reconfiguration of downstream institutions, infrastructures, value chains, social behaviours, political commitments, and so on. Unfortunately, however, this legitimation is premised on the idea of trying to do relatively little to support the reduction of GHG emissions, at a societal level, and finding the easiest material way to transform society.

In this chapter, then, I start with a conceptual discussion of the idea of political-economic materialities, drawing specifically on the work of Timothy Mitchell (2009, 2011). Afterwards, I turn to the promotion of advanced biofuels in Canada as an illustration of how these materialities not only constrain particular developments, discussed in the next chapter, but also *enable* the development of particular possibilities—in this case, drop-in biofuels.

NEOLIBERAL NATURES AND POLITICAL-ECONOMIC MATERIALITIES, OR MATERIAL POLITICAL ECONOMY

As I outlined in Chap. 2, there is a tendency in the neoliberal natures literature to assign some form of agency to 'nature' as an entity resisting, contesting, or challenging different forms of neoliberalization (Castree 2008a, b; Bakker 2009), ranging from the enclosure of nature (McCarthy and Prudham 2004) to its commodification (Roff 2008). On the one hand, this sort of approach can end up representing 'neoliberalism' as a singular process that is altered or hybridized as a result of its encounter with diverse biophysical restraints (Castree 2008a); and, on the other hand, it often presents 'nature' as an agent being acted on and acting upon neoliberalism, treating them both as distinct processes or systems. As such, the notion of neoliberal natures can, contradictorily, end up naturalizing both markets and natures—as separate and distinct categories and ontologies—rather than treating them as co-constructed; that is, there is no neoliberalism without nature, and no nature without neoliberalism (e.g. see Moore 2010). Taking the notion of co-construction seriously means, for me, treating nature-economy relations as always entangled and always implicated in each other's emergence and development; neoliberal natures—if we want to frame it as such—are therefore decidedly political-economic *and* material realities.

Contemporary carbon economies are material political economies. As Boykoff and Randalls (2009: 2299) argue, "carbon-based activities dominate [our] economies and societies in ways not seen before in human history"; as such, it makes sense to talk about a carbon economy in which human action, institutions, and infrastructures are entangled with the very materiality of natural and environmental processes relating to the discovery, extraction, processing, distribution, and consumption of carbon resources. Moreover, Timothy Mitchell (2011: 1) argues that we can also talk about 'carbon democracy'—the legitimation of fossil fuels as the underpinning for our societies—in which the materiality of "[f]ossil fuels helped to create both the possibility of modern democracy and its limits" (see also Mitchell 2009, 2010). In particular, Mitchell (2011: 7) notes that one of the key limitations represented by carbon energy—especially oil—"is that the political machinery that emerged to govern the age of fossil fuels, partly as a product of those forms of energy, may be incapable of addressing the events that will end it". I discuss this in the next chapter, but it's worth emphasizing that any bio-economy is likely to prove highly disruptive to the current carbon economy given the vastly different materialities between the two energy resources. It's hardly surprising, then, that recently there has been greater emphasis placed on the development of 'drop-in' biofuels, so named because of their ability to be used in existing distribution infrastructure and conversion devices with relatively few, if any, technical modifications compared to other biofuels.

To get to drop-in biofuels, however, means understanding the carbon economy. As a starting point, it's useful to turn to Mitchell's (2009, 2011) claim that oil-producing countries tend to be less democratic because they suffer from a 'resource curse'—more specifically, an 'oil curse'. According to Mitchell, the claim that countries suffer from an oil curse largely ignores "the ways oil is extracted, processed, shipped and consumed, the forms of agency and control these processes involve or the power of oil as a concentrated source of energy" (Mitchell 2009: 400). What Mitchell is getting at is that energy politics and the legitimation of energy regimes are bound up with the very materialities of energy itself, since these materialities shape the political forms, participation, and constraints of political activity. In addition to the simple biophysical characteristics of coal or oil, it's the material apparatus of energy production (e.g. mines), distribution (e.g. pipelines), and consumption (e.g. power stations) that shapes political power and, ultimately, the capacity and legitimacy for political and social change. Hence, it's critical to consider the materialities of biofuels and the bio-economy as an alternative energy regime to the fossil fuel regime that Mitchell concentrates on in his work.

The key to Mitchell's (2009, 2011) arguments is that the materialities of carbon energy (e.g. coal, oil) create possibilities for and limits on political action, which have to do with the biophysical characteristics of hydrocarbons themselves as well as the material (e.g. transport) and epistemic (e.g. accounting) apparatus needed to bring them into use. For example, coal enabled the rise of large-scale democratic movements (e.g. socialist political parties) driven by mass labour organizing in the nineteenth century, while oil, in contrast, reduced the capacity of workers to interrupt energy flows and exercise their collective power. All of this entailed a different political-economic materiality, involving a new technological and epistemic apparatus (see also Huber 2013). Ultimately, oil limited the development of democratic movements as it reduced the power of workers to back up their demands with acts of disruption. Mitchell highlights the necessary convergence of the materiality of oil with the epistemic practices and technologies of economics during the early years of the twentieth century. He also argues that the materiality of fossil fuels involves "the related networks, of international finance, for example, of technical knowledge, and of economic theory that different forms of energy depended upon and made possible" (Mitchell 2010: 190). This carbon economy—or, perhaps more precisely, 'carbon economics'—entailed a wholesale transformation of economics as a discipline, according to Mitchell (2011: Chap. 5), from a focus on natural resource depletion (regarding coal in the nineteenth century) to the treatment of oil (post World War II) as an 'inexhaustible resource' that reinforced the fiction of ever-rising national economic growth (also Boyer 2011). Here epistemic practices are entangled with the different materialities of coal and oil; the latter became bound up with new forms of national accounting (e.g. GNP), Keynesian demand management, and a focus on prices (i.e. 'petroknowledge') which presaged new economic technologies of calculation, price-setting, and so forth that "were built into the new financial institutions" (Mitchell 2011: 135).

I return to the importance of epistemic practices below when I consider the implications of the political materialities of bioenergy. I do so because, in accepting Mitchell's arguments, it's clear that an epistemic transition will be necessary with any material transition to a low- or zero-carbon economy. In coming back to how Mitchell relates to bioenergy, in general, it's crucial to consider his argument "that the political machinery that emerged to govern the age of fossil fuels, partly as a product of these forms of energy, may be incapable of addressing the events that will end it" (Mitchell 2011: 7). Or, more simply, we cannot rely upon a political apparatus underpinned by fossil fuels to engender and drive a systemic transition to a new energy regime based on renewables, including bioenergy and biofuels. New forms of energy entail new political legitimation machinery and new epistemic practices, which involve a completely different perspective to the carbon age. Any analysis of systemic material energy transitions, therefore, requires me to examine how that transition is legitimated and made politically and economically feasible.

A Bioenergy Regime?

I use bioenergy here to refer to the conversion of biomass from plants and waste streams into various forms of energy (e.g. electricity, heat) or energy carriers (liquid, gaseous, or solid fuels) (Calvert et al. 2017). As previous chapters illustrate, the last decade has been characterized by a significant push behind bioenergy and, specifically, liquid biofuels (e.g. bioethanol, biodiesel) as a key sustainability solution to climate change. Both bioenergy and biofuels are an important (and dominant) form of renewable energy in major world economies like the USA and EU. In the USA, for example, biofuels have a long history stretching back at least to the 1978 Energy Tax Act, which was concerned with US energy security following the oil crises in the 1970s (Kedron and Bagchi-Sen 2011). More recently, and as a result of the 2005 Energy Policy Act and 2007 Energy Independence and Security Act, the US overtook Brazil as the world's leading ethanol producer (Smith 2010). Thus it's no surprise that bioenergy now represents nearly half of the USA's renewable energy production (Zimmerer 2011). Similarly, in the EU, bioenergy represents a significant proportion of renewable energy production, over half in 2010 (ClientEarth 2012). Support for biofuels in the EU has been integrated into the 2003 Biofuels Directive and 2009 Renewable Energy Directive, while biofuels production has been primarily centred on biodiesel.

The dominance of bioenergy as a renewable energy source in the USA and EU reflects the policy support for biofuels in the transportation sector—a major greenhouse gas (GHG) emitter. As mentioned, this support dates back to the 1970s in some cases, largely as a response to the oil crises and fears about energy security (WorldWatch Institute 2007). The rationale behind promoting biofuels has since evolved a number of times in both the USA and EU; it has moved through several forms of policy legitimation including energy security, rural economic development, energy efficiency, and, finally, GHG emissions reductions following the 1997 Kyoto Protocol (e.g. Charles et al. 2007; Mol 2007). It's understandable that biofuels production has grown so significantly since 1990 (see Fig. 3.2), although this growth has been concentrated in the USA (ethanol) and EU (biodiesel) (Ponte 2014a).

Post-Kyoto, both the USA and EU began to articulate a sustainability rationale for legitimating biofuels in legislation like the 2000 US Biomass R&D Act and the 2003 EC Biofuels Directive (Charriere 2009; ClientEarth 2012). There are plenty of analyses of the positive and negative impacts of these pieces of legislation and later policy decisions, such as the 2006 EU Biofuels Strategy, 2002 US Farm Bill, and 2005 US Energy Policy Act (e.g. Charles et al. 2007; Londo and Deurwaarder 2007; McMichael 2009, 2012; Gillon 2010; Bailis and Baka 2011; Kedron and Bagchi-Sen 2011; Levidow et al. 2012a, b; Levidow and Papaioannou 2014). In the last few years, major media outlets have reported on the political and economic uncertainties surrounding biofuels, especially in terms of whether they will actually achieve their proposed environmental and socioeconomic benefits (Smith 2010). These uncertainties reflect growing criticism in the scientific literature about the ecological and social benefits of biofuels derived from primary agricultural products (e.g. corn, soy) and the land use changes these engender (Hill et al. 2006; Fargione et al. 2008; Mitchell 2008; Searchinger et al. 2008). Such criticisms largely focus on indirect land-use change (ILUC) as biofuels production in places like the US force changes in land-use in other parts of the world (cf. Harvey and Pilgrim 2011).

As a result of these concerns about conventional biofuels, there has been a significant policy push behind advanced biofuels, especially 'drop-in' versions. With advanced biofuels, net energy returns are meant to be greater and they are supposed to be derived from non-food crops (e.g. switchgrass, miscanthus) or biomass grown on non-agricultural land (e.g. forest residues) (Pimentel 2009; Sims et al. 2010; Stephen et al. 2011), and hence can be considered as more ecologically and socially sustainable (Bailis and Baka 2011; Levidow et al. 2012b). With 'drop-in' biofuels, chemical similarity to fossil fuel energy is meant to enable the direct substitution of petroleum/gasoline with advanced biofuels in automobiles, trucks, and airplanes, thereby reducing the overall costs of any low-carbon transition (Dale 2018). Given the uncertainties surrounding, and impediments to, the development and commercialization of these advanced drop-in biofuels (O'Connell and Haritos 2010; Tyner 2010a, b; Stephen et al. 2011), policy support is also increasing for research on new types of biofuels with higher energy contents (e.g. butanol). These so-called third or fourth generation biofuels are derived from algae or synthetic biology (Ferry et al. 2012), and are designed, again, to drop-in to prevailing infrastructures used by fossil fuels (Tyner 2010c; Savage 2011).

The prominence of bioenergy and support for drop-in biofuels as a key renewable energy resource in both the USA and EU has been reinforced recently by the 'bio-economy' strategies produced by these states in 2012 (e.g. CEC 2012; White House 2012)—also see Chaps. 4 and 6. Whether or not bioenergy and drop-in biofuels will or can engender a sustainable transition to a low- or zero-carbon future is open to question. It is my argument that whether this is likely depends upon the political-economic materialities of these bio-based energy regimes.

POLITICAL MATERIALITIES OF BIO-BASED ENERGY: THE CASE OF ONTARIO

Background to the Ontario Case Study

Building on Mitchell's (2009, 2010, 2011) arguments about the political materialities of carbon energy, my aim in this section is to apply his insights to bioenergy, especially the support for and development of advanced drop-in biofuels in the Province of Ontario, as well as Canada more generally. My interest in Ontario stems from the Provincial Government's ongoing and very active role in promoting sustainable energy transitions through various policies, which in many cases build on previous Federal Government policies (see Charriere 2009; Puddister et al. 2011; Mabee 2013). These policies include, but are not limited to:

- Ontario Provincial Government:
  - 2005 Ethanol Growth Fund
  - 2007 Ethanol in Gasoline Regulation
  - 2009 Ontario Green Energy Act

- Canada Federal Government:
  - 1995 Alternatives Fuels Act
  - 2000 Biomass for Energy Program
  - 2000 Action Plan on Climate Change

- 2003 Ethanol Expansion Program
- 2007 ecoENERGY for Renewable Power Initiative
- 2007 NextGen Biofuels Fund
- 2011 Federal Renewable Fuels Regulation

Ontario took a significant lead over other Canadian provinces when it comes to supporting bioenergy and biofuels (CanBio 2012). Until relatively recently, Ontario was heavily dependent upon fossil fuels for transport and electricity generation (Ontario Power Authority 2010). The transition to renewable energy, and especially bioenergy and biofuels, has been legitimated through the policies highlighted above.

The 2005 Ethanol Growth Fund (EGF) was established to support capital investment in ethanol production and assist producers in the face of market uncertainties, as well as fund R&D into biofuels. The EGF forms part of Ontario's plan to introduce a renewable fuel standard (RFS), which was enacted in the 2007 Ethanol in Gasoline Regulation (EGR). This was originally announced in 2004 as a provincial RFS, reflecting moves in other countries to introduce RFS based on biofuels (see Bailis and Baka 2011) and building on agreements with British Columbia and California to reduce GHG emissions (Charriere 2009). The emphasis on proportional mandates can be seen as part of a wider shift away from excise tax exemptions, which were highly variable between countries and even provinces (de Beer 2011). The EGR stipulated a five percent minimum ethanol blend by volume for all gasoline sold in Ontario from 2007—there's a similar Federal RFS introduced by the 2011 Renewable Fuels Regulation. The EGR has legitimated the creation of a market for over 880 million litres of ethanol for Ontario producers—a benefit of a RFS mandate over excise tax exemptions—most of which is produced from the conversion of starch from corn and wheat and is financially supported by the EGF. The investments through the EGF have resulted in the installation or construction of over 1000 Ml of ethanol production capacity in Ontario (Canadian Renewable Fuels Association 2011). According to de Beer (2011: 21), Ontario's EGR policy was at the time of its enactment unique because it contained (albeit in weak and, so far, ineffective form) provisions to encourage and support advanced biofuels production in Ontario, especially cellulosic biofuels from non-food crops (e.g. forests). The provincial government has also financially and politically legitimated commercial and pre-commercial advanced or drop-in biofuel production facilities through support for capital expenditures as well as licensing agreements on forest resources.

These policies in Ontario are representative of state interventions designed to promote and legitimate specific forms of sustainable transitions based on the (notional) de-carbonization of the economy. They imply not only a significant rethinking of the organization and configuration of energy production, distribution, and consumption, but also a rethinking of political formations and legitimation. Changes could be undertaken within the current energy regime by integrating renewables (e.g. bioenergy, biofuels) into prevailing infrastructures and institutions, or they could be pursued by totally disrupting the current energy regime. As Mitchell (2011) points out, the latter is considerably less likely because the 'political machinery' associated with fossil fuels is biased against the introduction of a new and competing energy regime. The main reason for this is that any new energy regime (e.g. bioenergy) entails new political machinery—that is, forms of legitimation—that necessarily contradicts the political machinery of previous energy regimes (e.g. fossil fuels). Moreover, any new political machinery will be tied to the biophysical and energetic qualities and characteristics of bioenergy and biofuels. This is especially the case if jurisdictions wish to achieve a scale of production that is meant to replace—in large part or in whole—our existing fossil-based energy regime (Richard 2010).

Legitimating What Biofuels?

The political changes associated with the transition to bioenergy and biofuels are frequently legitimated as socially, economically, and politically positive because they are expected to encourage things like local control and autonomy, decentralized decision-making and cohesion, and localized economic benefits like new jobs, new investment, and so on (Goldstein and Tyfield 2017). Deciding how biomass resources are best used is an inherently political choice with different political-economic implications in terms of end-users as well as patterns of production. If a given society chooses advanced drop-in biofuels, for instance, then production facilities will necessarily occur in large centralized facilities due to the need for economies of scale. However, if biomass is diverted instead towards combined heat and power or district heating systems, then it's more likely that a distributed pattern of development will occur because it's simply not possible to transfer heat over distances.

Each pathway entails different forms of legitimation and a 'new political machinery' that will be necessary to facilitate a transition toward bioenergy and biofuels. Here, I'm going to focus on three key issues with regard to the development of advanced biofuels in Ontario, bringing together in my analysis a consideration of the biophysical materialities of bioenergy and biofuels with how they enable a particular political machinery of legitimation. First, I discuss the implications of bio-based energy flows in order to analyse how different they are from fossil fuels and what their materialities end up legitimating. Second, I discuss the mobility of unprocessed biomass compared with fossil resources and what this, again, legitimates when it comes to the emerging political machinery of biofuels. Finally, I discuss the transboundary nature of bio-based energy in relation to legitimation concerns about sustainability and economic practices.

Bio-based Energy Flows
Mitchell's (2011: 12) concept of carbon democracy is based on 'buried sunshine' in the form of coal, oil, and gas. In contrast, bioenergy and biofuels can be considered 'grown sunshine', metaphorically speaking. When it comes to differentiating between the materiality of bio-based energy and carbon energy, it's notable that biomass has a relatively low energy density (i.e. GJ/t) compared to fossil energy resources and it grows at relatively fixed rates. On average, approximately 1.5 tonnes of biomass, grown aboveground, are required in order to replace the energy equivalent of one tonne of coal, which is recovered from subterranean deposits. Replacement values can be as high as 2–4 tonnes where oil and natural gas are concerned. Further, the rate at which biomass can be extracted from any given area must be limited in order to maintain ecological integrity at the site, including soil quality and niche habitats.

The low energy density of biomass legitimates the reduction of overall societal energy usage, if bio-based energy is going to entirely displace fossilized hydrocarbons. However, biomass couldn't possibly be used to power all sectors (e.g. heat, motor fuels, electricity) under existing rates and trends of global energy consumption—almost all estimates suggest that there just isn't enough solar energy being converted into biomass quickly enough, nor can biomass be extracted intensively enough, to allow that type of scenario to be sustainable (for a global-level perspective, see Berndes et al. 2003; for a local-level perspective, see Mabee and Mirck 2011). The shortfall grows larger in light of broader bio-economy strategies in which biomass is meant to replace oil as an input in the production of chemicals and plastics as well.

Even the most productive regions of the world will not produce enough biomass to support a bio-economy, so each society must greatly expand the land footprint of its energy system in order to realize the potential of bio-based energy. The land-intensive nature of bioenergy production will create and enable specific blockage points along bioenergy flows—like coal in Mitchell's (2011) argument—although the implications of this are uncertain. On the one hand, agriculture and forestry are sectors with low employment levels and traditionally low-levels of unionization, meaning that there is less likelihood of worker disruption. On the other hand, biomass will have to be grown, cut down, moved, processed, and refined in large quantities, meaning that there'll be plenty of blockage sites for disrupting these flows if workers so choose.

These three materialities of bio-based energy flows—low energy density, biomass conversion limits, and land footprint—help to legitimate certain political-economic developments and arrangements, namely, attempts to develop drop-in biofuels from algae and synthetic biology. As such, whether cellulosic biofuels are a failure or not (see Rapier 2018) can be treated as a side issue since another goal is replacing the production of cellulosic biofuels, primarily through the development of third- and fourth-generation biofuels that can simply replace fossil fuels by dropping-in to existing infrastructures (Savage 2011). As one informant noted, "the driver for drop-ins has always been the fact that there's less infrastructure required", although there are still issues since it might be "attractive on the one side [i.e. lower infrastructure costs] but has challenges in cost competitiveness on the other side" (Consultancy #123). However, the materialities of biomass might enable drop-in biofuels at the same time as they limit other bio-based production and products (e.g. chemicals). Another informant, for example, argued that:

> ...there's a big gap between the drop-in model, and the kind of bio-refinery model, where you are trying to take the feedstock and imagine it as several different products and not simply as an analogue for something that already exists. (Consultancy #135)

Biomass (Im)mobility

The second and related material characteristic of bio-based energy is that biomass is geographically distributed and relatively immobile. The low energy density (by weight and volume) of biomass means that it is not worthwhile in monetary or energetic terms to transport unprocessed biomass long distances from cultivation area to processing or refining plant (Hamelinck et al. 2005; Calvert et al. 2017). Biomass cultivation and processing activities must therefore occur at the same site or in sites very close to bioenergy production in order to achieve viable and relevant production scales. Furthermore, the procurement radius for a given facility and therefore the land-based transport requirements are generally much greater than fossil energy (remembering that bioenergy production scales with land area). Biomass co-firing projects in the USA, for instance:

> ...require supply chain managers to expand procurement from 2 to 3 coal suppliers supplying 16 million tonnes of coal to include 120 biomass suppliers supplying only 90,000 tonnes of biomass. (Wolf 2012: 46, citing Johnson 2012; see also Richard 2010)

At least three things matter here. First, the spread of biomass across a wide area and consequent spread of blockage points—see Mitchell (2011) on coal—mean that the power of workers to affect political change may be significantly curtailed as there will be numerous sources of inputs (e.g. biomass); as such, it would be relatively easy to shift from one sourcing site to another if biomass supplies are disrupted. However, at the same time, any biofuels production facility has less flexibility to switch suppliers because they cannot procure from a very long distance without incurring heavy economic costs. In other words, the friction of distance in biomass supply chains might bring some power balance between suppliers, producers, and workers. This highlights the crux of the chicken-and-egg situation that is stalling many biofuels investments; growers won't grow without a secure market and the market won't develop without a guarantee of a minimum supply at a fixed and acceptable price within a relatively small procurement radius. Second, and related to the last point, a range of local upstream actors (e.g. growers, land managers, biomass aggregators) must be coordinated long before and long after project implementation in order to secure the biomass necessary to keep a bio-based energy system operational. Finally, oil can be moved by pipeline and coal from mine-to-facility by rail, while biomass must be collected from a wide geographical area and trucked to a rail terminal or shipping port prior to bulk processing and/or transportation. This higher traffic activity associated with biomass transport and processing can be a source of local resistance to project development (Sampson et al. 2012).

Now, these materialities of biomass might lead to the assumption that biomass has to be consumed locally. For example, attempts to legitimate biofuels are liable to entail some concerns about how bio-based energy systems will impact local landscapes, in that the extraction, distribution, and conversion of energy will be more visible to a greater proportion of the population than is currently the case under a fossil energy regime (Calvert and Simandan 2010). Similarly, political-economic concerns reflect a similar focus on the impacts of localized biomass use, as these two informants emphasized:

> ...in the early years there's a lot of discussion around whatever region you were in in North America about using the local kind of biomass type feed-stock. (Biofuels Producer #118)

> ...it's the cost of logistics. So, the further you go from your facility the more expensive it is to get it to your plant. (Biofuels Producer #125)

However, this is not necessarily the only consequence of the immobility of biomass. Contradictorily perhaps, the upstream components of the bio-fuels value chain might be distributed, land intensive, and a potential drain on local biomass resources, but their very materialities—seemingly limited by local geographical dynamics—provide the driver for finding ways to process biomass into a densified energy carrier like wood pellets or liquid biofuels, since this makes it possible to distribute bio-based energy through global transportation networks and, thereby, extend the geographical reach of biomass supply chains.

As Mitchell (2011: 6) suggests "[C]itizens have developed ways of eating, travelling, housing themselves and consuming other goods and services that require very large amounts of energy from oil and other fossil fuels." On the one hand, replacing fossil fuels with bio-based energy will not, in and of itself, change these public and consumer preferences and expectations, meaning that biomass has to be *globalized* from the start. On the other hand, the materialities of biomass will shape how different publics engage with bio-based energy (Walker and Cass 2007); for example, there are already attempts to enrol local communities in decision-making processes as a way to legitimate renewable energy development (Yappa 2012), whether biofuels or otherwise. The globalization of biomass distribution necessarily entails this new political machinery of legitimation to resolve potential conflicts between the local production of biomass and its (potentially) global consumption as biofuels. Which side of the equation benefits most is a key question, and part of it revolves around transboundary issues, to which I now turn.

Sustainability and Transboundary Leakage

Materially, transboundary issues are central to any analysis of bio-based energy in a way that they are not when it comes to fossil energy—at least, when coal and oil regimes were first emerging. The transboundary nature of biofuels, for example, not only concerns the spatio-temporal aspects of biomass cultivation (e.g. daily or seasonal variability), there are also socio-material concerns (e.g. pricing or commodification variability). Generally, bio-based energy involves several overflows between material and political jurisdictions—national borders and geographical spaces (Giordano 2003). In particular, ClientEarth (2012: 16) identifies these (boundary) over-flows as involving 'geographical' and 'sectoral' loopholes in accounting for carbon emissions and emissions reductions from biofuels. Key here is understanding how the materiality of biofuels and the accounting of their sustainability legitimates particular forms of biomass trade.

First, socio-economic transboundary overflows include major differ-ences between different market development policies (MDPs), which I discuss in the next chapter. There is a significant difference, for example, between excise tax exemptions and RFS mandates, which makes the latter materially more attractive as a policy mechanism in promoting biofuels production. On the one hand, tax exemptions promote the transportation of bio-based energy products (e.g. ethanol) from low-cost producing regions (e.g. the US cornbelt) to areas where tax exemptions provide a market advantage (e.g. Ontario). On the other hand, RFS mandates are often supported by production incentives to develop local or domestic industry, which makes them more politically acceptable (Upham et al. 2007). This helps to explain why the Ontario Provincial Government removed the biofuels tax exemption and used the resultant tax revenue instead to fund local ethanol producers through the EGR, which meant that Ontario was no longer paying international producers.

Secondly, socio-material transboundary overflows cover the attribution of sustainability credits when it comes to accounting for the contribution of biofuels to GHG emissions reductions. Who gets to claim those cred-its? Is it the biomass producer, or the biofuels user? Capturing these energy flows and sustainability credits necessitates new forms of account-ing and calculation (i.e. political-economic technologies) which supersede those highlighted by Mitchell (2010, 2011) when it comes to the carbon economy. As such, bio-based energy is bound up with particular political-economic practices and expertise to account for things like transboundary overflows of sustainability benefits (Ponte 2014b). When it comes to bioenergy, there's a significant tension, which is bound up with new political-economic technologies (e.g. sustainability accounting) that enable some countries to claim credit for the sustainable benefits of bio-fuels (and other bio-based energy). In the Kyoto Protocol, for example, CO_2 emissions released by extracting biomass are assigned to the country of origin (i.e. producers) rather than combustion (i.e. users) (ClientEarth 2012). Consequently, where an Annex 1 country—basically the Global North—imports biomass from the Global South, they "are not held responsible for carbon dioxide emissions from bioenergy" (ibid.: 19). Annex 1 countries can report zero emissions from bioenergy use without evidence of actual emissions reduction, while the effects of biomass culti-vation, transportation, and export (e.g. land use change) are not accounted for in the assessment of their sustainability. This means that major econo-mies like the USA and EU can legitimate increasing their emissions as long as they import bioenergy from other places where any emissions are assigned, but not accounted for.

CONCLUSION

A sustainable transition is premised upon moving from a carbon energy regime to a renewable energy regime—a highly contested political-economic transformation, to say the least. In places like the USA and EU, the main form of renewable energy is bioenergy, especially biofuels. Recent policy and industry efforts are increasingly focusing on the development and implementation of what are known as 'drop-in' biofuels, so named because they can be incorporated into existing distribution infrastructure (e.g. pipelines) and conversion devices (e.g. combustion engine) with rela-tively few, if any, technical modifications. As with carbon energy, biofuels have particular materialities that are implicated in the political-economic possibilities and constraints facing societies around the world. These political-economic materialities of biofuels enable the pursuit of certain developments, as much as they constrain the development of others (see Chap. 6). In this chapter, I examined the political-economic materialities of biofuels as a way to understand how these materialities legitimate the development of certain forms of biofuels—namely, drop-in biofuels.

My focus on the promotion of a green or sustainable transition—in order to shift societies and economies away from dependence upon fossil fuels—provides the lens through which to examine particular technological pathways. Why are some chosen over others, for example? Often, such transitions are represented as an almost entirely positive transformation of a society towards low-or zero-carbon energy, jobs, and economies. However, analysing the political-economic materialities of biofuels raises several troubling issues. While the notion of a bio-economy—discussed throughout this book—is clearly legitimated on the basis of the sustainability benefits expected from the use of biomass rather than fossil fuels, and in some cases in terms of the expected new investments in rural or peripheral regions, there are real and perceived negative impacts associated with its implementation that, as discussed here, are directly related to the materialities of biomass and how these materialities impact bio-based and fossil-based energy supply chains as well as societal interaction with energy production, distribution, and use.

Attempts to integrate biofuels and other bioenergy into existing infrastructures and institutions—as drop-in substitutes—are likely to be problematic, not only because this will merely reinforce the prevailing and incumbent carbon economy but also because the materialities of biofuels will disrupt existing energy systems as well as regional economies, land-use systems, and transport infrastructure. The materiality of biomass necessarily problematizes the notion of drop-in biofuels, even though the development of the latter might be legitimated by that very materiality. Although these bio-based fuels have been processed (e.g. de-oxygenated, reformed into long carbon chains) to mimic carbon fuels, and therefore be compatible with existing infrastructure for fuel distribution (e.g. pipelines) and conversion (e.g. internal combustion engines), there are significant upstream changes and impacts as well. These include the way land is used and valued; where production facilities will be located; the sheer number and spatial distribution of resource (land) owners that must be considered; and increasing transportation requirements (e.g. for biomass). These things cannot be so easily 'dropped-in' to a carbon economy, nor are they the simple effect of specific materialities. As I show in Chap. 7, these nature-economy relations are co-constructed through the interaction between environmental and economic processes.

References

Bailis, R., Baka, J. (2011) Constructing sustainable biofuels: Governance of the emerging biofuel economy, *Annals of the Association of American Geographers* 101 (4): 827–838

Bakker, K. (2009) Commentary: Neoliberal nature, ecological fixes, and the pitfalls of comparative research, *Environment and Planning A* 41: 1781–1787.

Berndes, G., Hoogwijk, M. and van den Broek, R. (2003) The contribution of biomass in the future global energy supply: A review of 17 studies, *Biomass and Bioenergy* 25(1): 1–28.

Boyer, D. (2011) Energopolitics and the anthropology of energy, *Anthropology News* 52(5): 5–7.

Boykoff, M. and Randalls, S. (2009) Theorizing the carbon economy: Introduction to the special issue, *Environment and Planning A* 41(10): 2299–2304.

Calvert, K. and Simandan, D. (2010) Energy, space, and society: A reassessment of the changing landscape of energy production, distribution, and use, *Journal of Economics and Business Research* 16(1): 13–37.

Calvert, K., Birch, K. and Mabee, W. (2017) New perspectives on an old energy resource: Biomass and emerging bio-economies, in S. Bouzarovski, M. Pasqualetti and V. Castan Broto (eds) *The Routledge research companion to energy geographies*, London: Routledge, pp. 47–60.

Canadian Renewable Fuels Association. (2011) Plant locations (ethanol), *greenfuels.org*, retrieved from: http://www.greenfuels.org/en/industry-information/plants.aspx.

CanBio. (2012) *Economic impact of bioenergy in Canada – 2011*, Ottawa: Canadian Bioenergy Association, retrieved from: http://www.canbio.ca/upload/documents/canbio-bioenergydata-study-2011-jan-31a-2012.pdf.

Castree, N. (2008a) Neoliberalising nature: The logics of deregulation and rereg-ulation. *Environment and Planning A* 40: 131–152.

Castree, N. (2008b) Neoliberalising nature: Processes, effects, and evaluations, *Environment and Planning A* 40: 153–173.

CEC. (2012) *Innovating for sustainable growth: A bioeconomy for Europe* [COM(2012) 60 Final], Brussels: Commission of the European Communities.

Charles, M., Ryan, R., Ryan, N. and Oloruntoba, R. (2007) Public policy and biofuels: The way forward? *Energy Policy* 35(11): 5737–5746.

Charriere, A. (2009) *Bioenergy policy and regulation in Canada*, Ottawa: Regulatory Governance Initiative, Carleton University, retrieved from: http://www.regulatorygovernance.ca/publication/timeline-bioenergy-policy-and-regulation-in-canada/wppa_open/.

ClientEarth. (2012) *Carbon impacts of bioenergy under European and interna-tional rules*. Brussels: ClientEarth.

Dale, B. (2018) Time to rethink cellulosic biofuels? *Biofuels, Bioproducts & Biorefining* 12(1): 5–7.

Dahmann, K., Fowler, L. and Smith, P. (2016) United States law and policy and the biofuel industry, in Y. Le Bouthillier, A. Cowie, P. Martin, and H. McLeod-Kilmurray (eds) *The law and policy of biofuels*, Cheltenham: Edward Elgar, pp. 102–140.

de Beer, J. (2011) *Network governance of biofuels*, Valgen Working Paper Series, retrieved from: http://jeremydebeer.ca/wp-content/uploads/2012/02/Network_Governance_of_Biofuels%20VALGEN%20Working%20Paper.pdf.

Fargione, J., Hill, J., Tilman, D., Polasky, S. and Hawthorne, P. (2008) Land clearing and the biofuel carbon debt, *Science* 319: 1235–1238.

Ferry, M., Hasty, J. and Cookson, N.A. (2012) Synthetic biology approaches to biofuel production, *Biofuels* 3(1): 9–12.

Gillon, S. (2010) Fields of dreams: Negotiating an ethanol agenda in the Midwest United States, *Journal of Peasant Studies* 37(4): 723–748.

Giordano, M. (2003) The geography of the commons: The role of scale and space, *Annals of the Association of American Geographers* 93(2): 365–375.

Goldstein, J. and Tyfield, D. (2017) Green Keynesianism: Bringing the entrepreneurial state back in(to question)? *Science as Culture* https://doi.org/10.1108 0/09505431.2017.1346598.

Hamelinck, C., Suurs, R. and Faaij, A. (2005) International bioenergy transport costs and energy balance, *Biomass and Bioenergy* 29(2): 114–134.

Harvey, M. and Pilgrim, S. (2011) The new competition for land: Food, energy, and climate change, *Food Policy* 36(1): 540–551.

Hill, J., Nelson, E., Tilman, D., Polasky, S. and Tiffany, D. (2006) Environmental, economic, and energetic costs and benefits of biodiesel and ethanol biofuels, *PNAS* 103(30): 11206–11210.

Huber, M. (2013) *Lifeblood: Oil, freedom, and the forces of capital*, Minneapolis: University of Minnesota Press.

Johnson, B. (2012) Biomass power: Planning, technical issues, and economic drivers, presented at *Biomass Power: Planning, Technical Issues, and Economic Drivers Conference*, Toronto, Canada, 22–23 February.

Kedron, P. and Bagchi-Sen, S. (2011) A study of the emerging renewable energy sector within Iowa, *Annals of the Association of American Geographers* 101(4): 882–896.

Levidow, L., Birch, K. and Papaioannou, T. (2012a) EU agri-innovation policy: Two contending visions of the knowledge-based bio-economy, *Critical Policy Studies* 6(1): 40–65.

Levidow, L., Papaioannou, T. and Birch, K. (2012b) Neoliberalising technoscience and environment: EU policy for competitive, sustainable biofuels, in L. Pellizzoni and M. Ylönen (eds) *Neoliberalism and technoscience: Critical assessments*, Farnham: Ashgate Publishers, 159–186.

Levidow, L. and Papaioannou, T. (2014) UK biofuel policy: Envisaging sustainable biofuels, shaping institutions, *Environment and Planning A* 46(2): 280–298.

Londo, M. and Deurwaarder, E. (2007) Developments in EU biofuels policy related to sustainability issues: Overview and outlook, *Biofuels, Bioproducts and Biorefining* 1(4): 292–302.

Mabee, W. (2013) Progress in the Canadian biorefining sector, *Biofuels* 4(4): 437–452.

Mabee, W. and Mirck, J. (2011) A regional evaluation of potential bioenergy production pathways in eastern Ontario, Canada, *Annals of the Association of American Geographers* 101(4): 897–906.

McCarthy, J. and Prudham, S. (2004) Neoliberal nature and the nature of neoliberalism, *Geoforum* 35: 275–283.

McMichael, P. (2009) The agrofuels project at large, *Critical Sociology* 35(6): 825–839.

McMichael, P. (2012) The land grab and corporate food regime restructuring, *Journal of Peasant Studies* 39(3–4): 681–701.

Mitchell, D. (2008) *A note on rising food prices*, Washington, DC: World Bank.

Mitchell, T. (2009) Carbon democracy, *Economy and Society* 38(3): 399–432.

Mitchell, T. (2010) The resources of economics: Making the 1973 oil crisis, *Journal of Cultural Economy* 3(2): 189–204.

Mitchell, T. (2011) *Carbon democracy*, London: Verso.

Mol, A. (2007) Boundless biofuels? Between environmental sustainability and vulnerability, *Sociologia Ruralis* 47(4): 297–315.

Moore, J.W. (2010) The end of the road? Agricultural revolutions in the capitalist world-ecology, 1450–2010, *Journal of Agrarian Change* 10(3): 389–413.

O'Connell, D. and Haritos, V. (2010) Conceptual investment framework for biofuels and biorefineries research and development, *Biofuels* 1(1): 201–216.

Ontario Power Authority (OPA). (2010) *Ontario's long-term energy plan*, Toronto: Ontario Power Authority.

Pimentel, D. (2009) Biofuel food disasters and cellulosic ethanol problems, *Bulletin of Science, Technology and Society* 29(3): 205–212.

Ponte, S. (2014a) The evolutionary dynamics of biofuel value chains: From unipolar and government-driven to multipolar governance, *Environment and Planning A* 46(2): 353–372.

Ponte, S. (2014b) "Roundtabling" sustainability: Lessons from the biofuel industry, *Geoforum* 54: 261–271.

Power, M. (2016) Lessons from US biofuels policy: The renewable fuel Standard's rocky ride, in Y. Le Bouthillier, A. Cowie, P. Martin, and H. McLeod-Kilmurray (eds) *The law and policy of biofuels*, Cheltenham: Edward Elgar, pp. 141–163.

Puddister, D., Dominy, S., Baker, J., Morris, D., Maure, J., Rice, J., Jones, T., Majumdar, I., Hazlett, P., Titus, B., Fleming, R. and Wetzel, S. (2011) Opportunities and challenges for Ontario's forest bioeconomy, *The Forestry Chronicle* 87(4): 468–477.

Rapier, R. (2018) Cellulosic ethanol falling far short of the hype, *forbes.com*, 11 February, retrieved from: https://www.forbes.com/sites/rrapier/2018/02/11/cellulosic-ethanol-falling-far-short-of-the-hype/2/#3d0c613d6a6e

Richard, T. (2010) Challenges in scaling up biofuels infrastructure, *Science* 329: 793–796.

Roff, R.J. (2008) Preempting to nothing: Neoliberalism and the fight to de/re-regulate agricultural biotechnology, *Geoforum* 39: 1423–1438.

Sampson, C., Agnew, J. and Wassermann, J. (2012) *Logistics of agricultural-based biomass feedstock for Saskatchewan*, Report prepared for ABC Steering Committee, SaskPower, NRCan. Project No. E7810.

Savage, N. (2011) The ideal biofuel, *Nature* 474: S9–S11.

Schnepf, R. and Yacobucci, B. (2013) *Renewable Fuel Standard (RFS): Overview and issues*, Congressional Research Service, 7-5700, retrieved from: https://www.ifdaonline.org/IFDA/media/IFDA/GR/CRS-RFS-Overview-Issues.pdf.

Searchinger, T., Heimlich, R., Houghton, R., Dong, F., Elobeid, A., Fabiosa, J., Tokgoz, S., Hayes, D. and Yu, T.-H. (2008) Use of US croplands for biofuels increases greenhouse gases through emissions from land-use change, *Science* 319: 1238–1240.

Sims, R., Mabee, W., Saddler, J. and Taylor, M. (2010) An overview of second generation biofuel technologies, *Bioresource Technology* 101(6): 1570–1580.

Smith, J. (2010) *Biofuels and the globalization of risk*, London: Zed Books.

Stephen, J., Mabee, W. and Saddler, J. (2011) Will second-generation ethanol be able to compete with first-generation ethanol? Opportunities for cost reduction, *Biofuels, Bioproducts & Biorefining* 6(2): 159–176.

Tyner, W. (2010a) Cellulosic biofuels market uncertainties and government policy, *Biofuels* 1(3): 389–391.

Tyner, W. (2010b) Comparisons of the US and EU approaches to stimulating biofuels, *Biofuels* 1(1): 19–21.

Tyner, W. (2010c) Why the push for drop-in biofuels? *Biofuels* 1(6): 813–814.

Upham, P., Shackley, S. and Waterman, H. (2007) Public and stakeholder perceptions of 2030 bioenergy scenarios for the Yorkshire and Humber region, *Energy Policy* 35(9): 4403–4412.

Walker, G. and Cass, N. (2007) Carbon reduction, "the public" and renewable energy: Engaging with socio-technical configurations, *Area* 39(4): 458–469.

White House. (2012) *National bioeconomy blueprint*, Washington, DC: The White House.

Wolf, D. (2012) *Adjusting expectations of scale based on limitations of supply: A review of the case for a forest bioenergy strategy that prioritizes decentralization, efficiency, and integration*, MSc Thesis in Forestry, Faculty of Forestry, University of Toronto.

Worldwatch Institute. (2007) *Biofuels for transport*, London: Earthscan.

Yappa, R. (2012) Germany's renewable powerhouse: City utilities spearhead switch to renewables, *Renewable Energy World* September–October 2012: 48–53.

Zimmerer, K. (2011) New geographies of energy: Introduction to the special issue, *Annals of the Association of American Geographers* 101(4): 705–711.

CHAPTER 6

Material Limits to Bio-Economies

INTRODUCTION

As I've stressed already in this book, countries like Canada urgently need to find ways to transition from a fossil fuel energy system to a renewable and sustainable energy system if they are to ameliorate the worsening effects of climate change. Numerous technologies are offered as solutions towards this end, from carbon capture and storage through geoengineering to liquid biofuels, the latter being the focus of my own research on which this book is based. While there are many technological possibilities to support these kinds of socio-technical or sustainability transitions, it is important to examine *how* such transitions happen and *how* they are supported or derailed. A growing literature on these sorts of sustainability transitions does just that, as discussed in Chap. 2 (e.g. Geels 2002, 2004, 2005; Geels and Schot 2007; Coenen et al. 2012; Tyfield 2014; Hansen and Coenen 2015).

For a variety of reasons, which I discuss in other chapters in the book, liquid biofuels (henceforth biofuels) have become a particularly popular socio-technical transition pathway in countries like Canada—as well as the USA and EU. First generation, or conventional, biofuels are derived from food crops like sugarcane, corn, canola, and jatropha; however, these conventional biofuels have been criticized for their impacts on and implications for global food security and for their failure to reduce greenhouse gas emissions (e.g. Fargione et al. 2008; Searchinger et al. 2008). As a result, many countries have sought to promote second generation, or

© The Author(s) 2019
K. Birch, *Neoliberal Bio-Economies?*,
https://doi.org/10.1007/978-3-319-91424-4_6

advanced, biofuels, which are derived from non-food crops and biomass like wood, switchgrass, agricultural or forestry residues, waste, and simple organisms (e.g. algae). Advanced biofuels are meant to solve the negative effects of first generation biofuels by providing a higher energy return on biomass use and by not competing with food crops (Mabee 2007; Sims et al. 2010; Puddister et al. 2011). In particular, advanced biofuels hold considerable potential for countries like Canada that have a significant biomass resource base (Mabee et al. 2005; Kedron 2015), although it's important to remember that even though Canada is in the top six ethanol producers, it still only produces 1.9 percent of global ethanol (in 2014).

Over the last few years, conventional and advanced biofuels have been integrated into a broader vision of societal transformation called the bio-economy (e.g. CEC 2005, 2012; OECD 2006; White House 2012). As discussed in the previous chapters, the bio-economy is envisioned as a key transition pathway from an economy dependent on a finite, polluting resource (i.e. fossil fuels) to an economy dependent on renewable, clean resources—basically, biological plant matter (Pellerin and Taylor 2008; Frow et al. 2009; Mathews 2009; Birch et al. 2010; Scarlat and Dallemand 2011; Shortall et al. 2015). Broader than (bio-)energy, the bio-economy represents a more fundamental transformation of society, economy, and culture (cf. Geels 2002). However, it's notable that the bio-economy retains the particular focus in the sustainability transition literature on changes *within* techno-economic systems—or 'systems renovation'—rather than a wholesale transformation *of* those same systems themselves (Tyfield 2017).

By this, I mean that much of the transition literature, from whatever theoretical angle, is compatible with the dominant political-economic discourse, practices, and system increasingly characterized as 'neoliberalism' (see Chap. 2). As with most new technologies, for example, the development of advanced biofuels is driven by prevailing social, institutional, and infrastructure pressures and path dependencies like consumer and market demand (Geels 2005; Schot and Geels 2008); it's not being driven by the idea of wholesale political-economic transformation. As such, it's useful to look at how particular transition pathways are rolled out as solutions to climate change in order to understand how transition pathways like the bio-economy are underpinned by specific techno-economic assumptions (Birch 2017). It's also important, here, to understand the forms of resistance and contestation to these assumptions, especially as embodied in the biophysical materialities of 'nature' itself—a key claim made in the neoliberal natures literature.

In order to do this, I examine the deployment of 'market development' policies (or MDPs) to support the emergence, development, and expansion of the bio-economy around the world and in Canada—using advanced biofuels as a case study. These MDPs construct new markets through labelling, standard setting, mandates, supply chain configurations, and other actions. As an example of how new markets are constructed, this chapter illustrates the extent to which nature-economy relations are configured and mediated by state actors—a claim made by critics of neoliberalism (e.g. Peck 2010; Mirowski 2013; Brown 2015)—and non-state actors through the establishment of monopolies, contractual arrangements, and other mechanisms that conflict with notions of 'free' markets. As such, I aim to problematize both the claims about the potential of sustainability transitions *and* the claims that nature is increasingly 'neoliberalized' through the extension of markets and market relations (see Chap. 2).

NEOLIBERALISM, MARKET DEVELOPMENT, AND SOCIETAL TRANSITIONS

Neoliberalism is defined as the installation of markets as *the* foundational societal institution, to the exclusion of other institutions like the family, state, national or cultural identity, or otherwise. As noted in Chap. 2, there's a burgeoning literature on the neoliberalization of nature, or 'neoliberal natures' (e.g. Castree 2008a, b; Bakker 2009, 2010; Collard et al. 2016). This work is premised on the idea that environmental issues are increasingly approached from a market-based perspective, and market-based instruments or mechanisms are increasingly used to solve environmental problems like biodiversity loss, water and air pollution, environmental degradation, and so on. For example, a key idea—which goes back to the work of Garrett Hardin (1968) on the so-called tragedy of the commons—is that environmental issues often result from a lack of clear property rights because people over-use things they can access freely (e.g. common grazing land), but do not own (Birch et al. 2017). As a result of these sort of ideas, private ownership has been extended to more and more natural phenomena, such as water or the atmosphere (Loftus and Budds 2016; Lohmann 2016).

While the neoliberal natures literature provides important insights into various attempts to address environmental problems, it's premised on a number of assumptions. In particular, critics of neoliberalism tend to claim that markets are both an aberration—in that they are imposed on society or nature—*and* a form of 'instituted process' (see Polanyi 1957), in that they are social constructs rather than the result of ahistorical economic laws (a la neoclassical economics). Taking the latter point seriously, as I wish to do here, means thinking about how markets are created, extended, and supported through deliberate action, *and*, as importantly, how 'market development' is embedded within nature-economy relations, rather than disembedded from those relations. For example, it's important to think about how biophysical materialities both enable *and* constrain what types of market can and do emerge (and vice versa), a discussion I come back to more fully in Chap. 7. Such issues are integral to broader debates about sustainability or socio-technical transitions, in that these latter debates are concerned with the development and emergence of new (socio-technical) markets and political-economic systems.

Current research on sustainability transitions focuses on how prevailing socio-technical 'regimes' change as a result of emerging technological 'niches'—see Geels (2005), for example, on the shift from horse-drawn transport to the automobile. As with most socio-technical transformations, however, transitions are enabled and constrained by prevailing social, institutional, and infrastructural pressures and path dependencies (Geels 2005; Schot and Geels 2008). Here, I mention two examples relevant to liquid biofuels. First, advanced biofuels are increasingly designed as 'drop-in' fuels that can be integrated into existing socio-technical systems and markets (Birch and Calvert 2015), thereby avoiding the costly reconfiguration of transport and energy infrastructure, business supply chains, and market institutions. I outline this in more detail in Chap. 5. For now, however, it is notable that drop-in biofuels end up reinforcing the carbon lock-in of societies to existing energy systems and markets underpinning fossil fuel economies (e.g. biofuels are designed so that there is no reason to rethink automobile use). Second, the development of new technologies is highly uncertain in that they "cannot immediately compete on the market against established technologies" (Schot and Geels 2008: 537). At issue here is the fact that advanced biofuels have to emerge *within* prevailing socio-technical systems and markets where they threaten incumbent industries (e.g. petroleum, corn ethanol). From the perspective that markets are *instituted* rather than naturalistic (see above), this means that it's crucial to analyse how markets for new technologies are created by *market development* policies (e.g. mandates, subsidies, standards, regulations, etc.) or stymied by dominant, incumbent social actors. To the extent that broad systemic change is required for sustainability transitions, the development of advanced biofuels could very easily be derailed by both infrastructure lock-in and incumbent market power.

Something is missing in this analysis, however. And that is an understanding of sustainability transitions as both socio-technical *and* material processes. As I've argued elsewhere (e.g. Birch 2013, 2016a, b), it's necessary to analyse the particular materialities of different transition pathways, like the bio-economy, in order to analyse how different biophysical characteristics enable, limit, and/or (re)configure specific political, economic, social, and technological transformations. By this, I refer to what Geels (2005: 451) calls the "material aspects of society"—or, "material and spatial arrangements" defined as the socio-technical 'landscape'—and their influence on prevailing socio-technical regimes and emerging niches. It is for this reason that I draw on ideas from environmental economic geography (EEG)—see Chap. 2—and the analytical emphasis on taking environmental *materiality* into account when analysing economic processes (e.g. Bridge 2008; Hudson 2008). Although Geels (2005) does highlight the importance of the material 'landscape' in sustainability transitions, this misses the point that *all* regimes, niches, regime elements, niche elements, and so on have a materiality to them. For example, human bodies, organizations, social interaction, regulatory rules, and markets have particular biophysical characteristics that need to be taken into account. Attempts to create new markets are necessarily configured by these materialities in diverse ways, some enable particular markets and others constrain them (Birch 2013, 2016a; Birch and Calvert 2015).

My overall conceptual aim in the rest of this chapter is to analyse the 'market development' policies—by which I mean the subsidies, mandates, standards, labelling, contracts, and so on—used to promote the bio-economy, using advanced biofuels as my case study (see Daemmrich 2015 on bioplastics). Understanding these market development policies is important for understanding sustainability transitions since they combine both future visions *and* socio-material configurations. On the one hand, future visions are centrally implicated in promoting niche innovations which threaten the prevailing market actors and elements in existing socio-technical regimes (Geels 2005; Schot and Geels 2008; Bauer 2018)—see Chap. 4. And, on the other hand, while existing socio-technical regimes reflect specific socio-material configurations (e.g. cars, roads, oil, suburbs), these configurations simultaneously enable and constrain the development of new markets and sectors (e.g. biofuels). However, taking materiality seriously means trying to unpack regimes and niches as necessarily emergent from existing material landscapes as well as the specific materialities of their particular configuration. As others argue elsewhere, for example, the materiality of bioenergy and biofuels—their biophysical characteristics—is an essential factor in any understanding of their (market) development and political acceptance (Birch and Calvert 2015).

BIO-ECONOMY POLICY STRATEGIES AND MARKET DEVELOPMENT POLICIES AROUND THE WORLD

I examine the policy visions and imaginaries that underpin particular policy frameworks in Chap. 4. Here, I analyse the enactment of these visions and frameworks through the development of bio-economy policy strategies and the implementation of market development policies (MDPs) to promote the bio-economy. I start by outlining the development of bio-economy strategies around the world and then in Canada, more specifically. This leads into a discussion of the MDPs deployed in different political-economic jurisdictions and the identification of four key MDPs—mandates, subsidies, feedstock supply, and standards—when it comes to the promotion and development of conventional and advanced biofuels. I then finish the analysis by examining the specific implementation of these policies in the development of advanced biofuels in the Canadian context and consider the material limits that have constrained their implementation.

Bio-Economy Policy Strategies Around the World

By and large, the 'bio-economy' is an elite policy term. So far, it has little resonance with societal publics, wherever we might be in the world, perhaps even coming across as yet another piece of policy wonkery or jargon. It's primarily used to describe and outline a particular policy strategy for supporting the transition of societies from fossil fuel dependence towards the use of supposedly low-carbon biomass resources, like food and non-food crops, forests, aquaculture, waste, and, increasingly, synthetic biological matter. It is, in this regard, a concept tied, firstly, to a particular vision of a low-carbon future in which biological materials have replaced carbon materials as the underlying resource base and, secondly, to the idea that achieving this vision necessitates policy action by governments. Examples illustrating this dynamic abound. For instance, an editorial in the academic journal *Biofuels, Bioproducts and Biorefining* states that "The necessary transition to a sustainable bioeconomy will not happen overnight and may never happen without supportive governmental policies" (Chisti 2010: 361). It's rarely, if ever, associated with the idea of withdrawing policy and government intervention in markets, a simplistic notion of neo-liberalism that is trotted out in (usually popular) debates every so often.

As a concept, the bio-economy (and 'bio-based economy') has become a focus of growing academic attention, as Fig. 6.1 illustrates, and of political and activist concern (e.g. TNI 2015), business interest (e.g. Aguilar et al. 2018), and global policy-making (e.g. El-Chickakli et al. 2016). Several recent reviews of the academic literature highlight the growing interest in the area, as well as the continuing ambiguity surrounding the concept (e.g. Pfau et al. 2014; Golembiewski et al. 2015; Bugge et al. 2016). For example, as Bugge et al. (2016) note, the growing interest in the bio-economy has not necessarily clarified what people mean by the term, nor has it created consensus around a single concept.

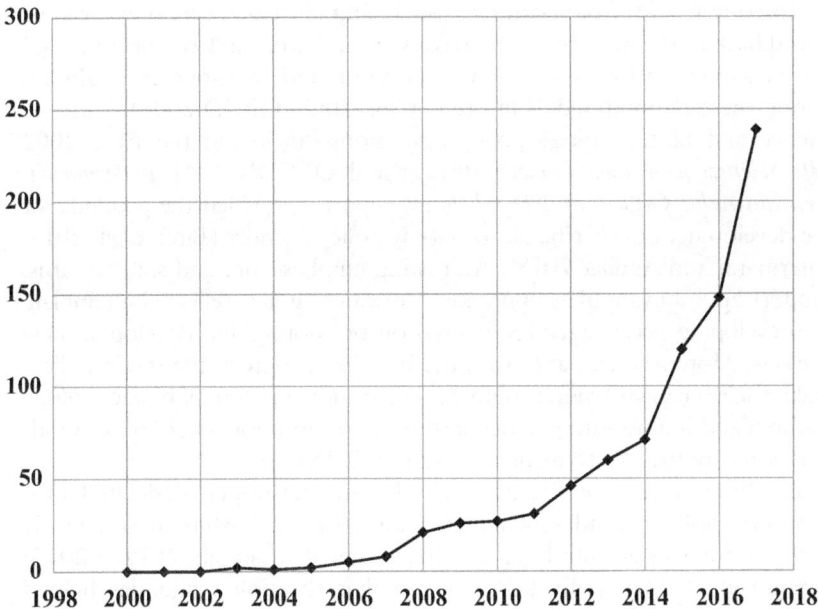

Fig. 6.1 Academic literature on bio-economy, 2000–2017. (Source: Web of Science; search terms "bioeconomy OR bio-economy" (total number = 807))

In part, the flexibility of the term 'bio-economy'—or the related term 'bio-based economy'—likely reflects a key reason for its popularity in policy circles, in that it can mean something to and for everyone (Frow et al. 2009; Richardson 2012; Mukhtarov et al. 2017). For business, it can mean new markets; for policy-makers and politicians, it can mean regional or rural development and new jobs; and for environmentalists, it can mean the reduction of greenhouse gas (GHG) emissions and end of fossil fuel dependency. Moreover, its flexibility means that the bio-economy can be deployed across a range of societal problems and global challenges, representing a potential solution to policy issues as wide-ranging as global hunger, energy security, sustainable economic growth and consumption, and the protection of biodiversity (e.g. El-Chickakli et al. 2016). In light of all this, it's perhaps helpful to try and understand where the bio-economy concept came from, at least as a way to unpack its current policy relevance. I've outlined these policy developments in Fig. 6.2, which represents a timeline of the development of bio-economy policy strategies in Europe (top) and the rest of the world (bottom).

The origins of the bio-economy—including bio-based economy—can be traced back to the work of policy-makers in the European Commission (EC) and Organisation for Economic Co-operation and Development (OECD) during the early- and mid-2000s (e.g. CEC 2005; OECD 2006). Early life science and biotechnology policy discussions—including the EC's 2002 *Life Sciences and Biotechnology Strategy* and OECD's 2004 *Biotechnology for sustainable Growth and Development* report—provided the grounds for the development of specific bio-economy policy agendas (Birch et al. 2014; Patermann and Aguilar 2018). A growing emphasis on, and shift towards, supporting 'innovation' as opposed to 'research' policy reflected a template for developing policy agendas focused on promoting and developing new markets. Moreover, and at least in the European context, the threat policy-makers felt the GMO moratorium could have for agricultural biotechnology research and innovation provided a stimulus to push for other uses of modern biotechnology (Patermann and Aguilar 2018).

By 2005, both the EC and OECD had developed dedicated bio-economy policy agendas, which fed into the establishment of specific policy frameworks (Birch et al. 2010; Levidow et al. 2012a, 2013; Schmid et al. 2012). By this, I mean that the policy agendas helped shaped a range of policy visions, instruments, organizational structures, and so on. The EC's bio-economy agenda was incorporated into the European Union's (EU) research and innovation funding programmes.

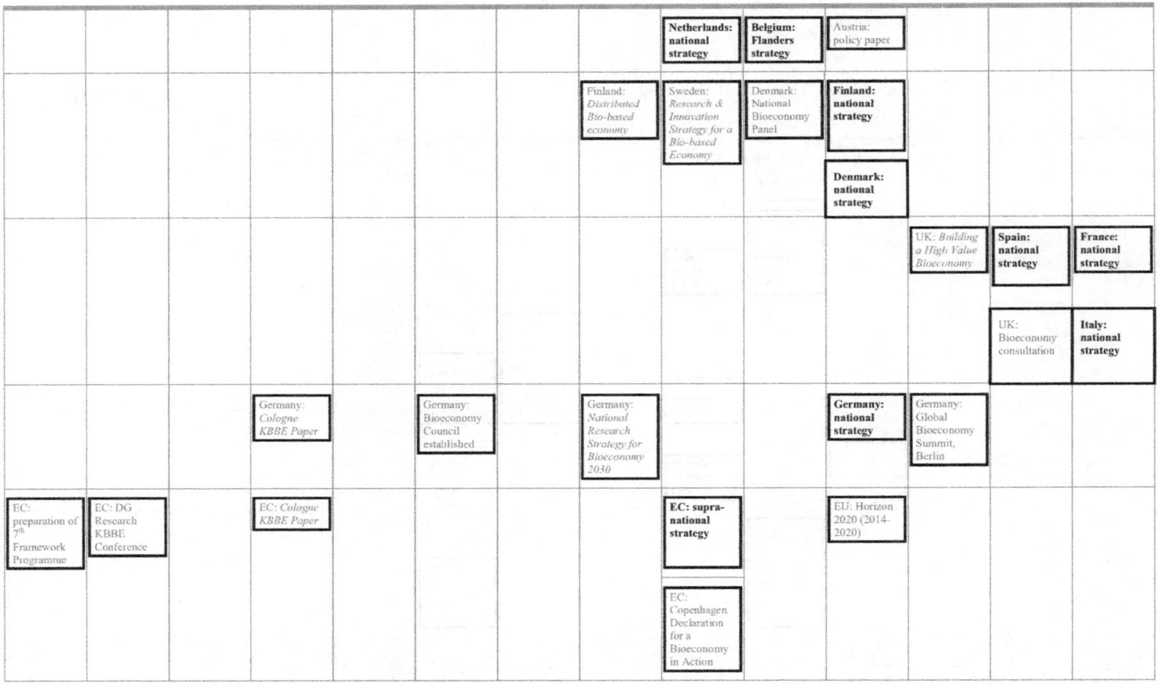

Fig. 6.2 Bio-economy policy timeline. (Source: Germany Bioeconomy Council, BioSteps, European Commission, https://biobs.jrc.ec.europa.eu/)

	2004	2005	2006	2007	2008	2009	2010	2011	2012	2013	2014	2015	2016	2017
	OECD: *Biotechnology for sustainable Growth and Development*		OECD: *Bioeconomy to 2030: Designing a Policy Agenda* [Proposal]			OECD: *Bioeconomy to 2030: Designing a Policy Agenda* [Findings]								
						Canada: BioteCanada *Beyond Moose & Mountains*				Canada: federal government establishes Bioeconomy Interdepartmental Working Group				
									USA: national strategy		USA: BioPreferred program expanded			
						Japan: Basic Act for Promotion of Biomass Utilization	Japan: National Biomass Policy Council & National Plan for Biomass Utilization		**Japan: national strategy**					
									Russia: national strategy	**South Africa: national strategy**				
									Malaysia: national strategy					

Fig. 6.2 (continued)

For example, it was incorporated into the 7th Framework Programme (FP7) (2007–2013) and Horizon 2020 (2014–2020) through the establishment of several dedicated 'European Technology Platforms', or ETPs (see Levidow et al. 2012a, b)—see Chap. 8. Several ETPs are included under the bio-economy umbrella, including the following:

- European Aquaculture Technology and Innovation Platform
- European Technology Platform for Global Animal Health
- Farm Animal Breeding and Reproduction Technology Platform
- Forest-bases Sector Technology Platform
- Food for Life
- Plants for the Future
- (and after some debate) Technology Platform Organics

Some ETPs—like the Innovative Medicines Initiative (IMI) established in 2008—have been formalized as public-private partnerships called Joint Technology Initiatives (JTI) (Birch et al. 2014). However these various ETPs are organized, their purpose is to develop and coordinate policy agendas, especially in terms of focusing research and innovation priorities and then enrolling public and private stakeholders in support of those priorities.

Alongside these sort of initiatives, a growing number of countries and other jurisdictions around the world have created dedicated bio-economy policy strategies designed to support the development of bio-based sectors, including biofuels, bioplastics, biochemicals, and so on (McCormick and Kautto 2013; Staffas et al. 2013; McCormick 2014; de Besi and McCormick 2015; Hausknost et al. 2017) As Fig. 6.2 illustrates, these strategies are relatively new, most being established since 2012, at least as formal 'strategies'. Policy developments in each country are obviously distinct, but it is possible to trace the development of these strategies in several cases. For example, Germany's 2014 bio-economy strategy can be traced back to the earlier 2011 *National Research Strategy for Bioeconomy 2030*, as well as the 2009 establishment of the German Bioeconomy Council (or Bioökonomierat) and 2007 'Cologne Paper' on the bio-economy produced while Germany held the EU presidency. Germany's bio-economy strategy is very much government-driven and oriented around creating the national research and economic conditions needed to support the bio-economy. In contrast, places like Canada have no national bio-economy policy strategy (see Chap. 4), and any proposals that have emerged are primarily business-driven (Birch 2016b). Examples here include the 2009 BIOTECanada 'blueprint' called *Beyond Moose and Mountains* and the 2014 Canadian Renewable Fuels Association 'action plan' called *Evolution & Growth: From Biofuels to Bioeconomy*. For any reader interested in exploring these policy strategies in more detail, the best place to start is the Germany Bioeconomy Council (see German Bioeconomy Council 2015a, b).

Market Development Policies in the Bio-Economy Around the World

An important aspect of these various bio-economy strategies is how they try to implement what can sometimes seem like high-minded or utopian policy visions and agendas into on-the-ground action. My particular focus in this chapter is on the deployment of MDPs to support biofuels, biomaterial, and biochemical markets. A range of academic and policy literature highlights the variety of market development policies—or MDPs—that have been tried out around the world (e.g. Sorda et al. 2010; German Bioeconomy Council 2015a, b; Scarlat et al. 2015), especially in relation to advanced biofuels. I've outlined the various MDPs in more detail and included examples in Table 6.1 below. These MDPs cover: *mandates* and *subsidies* to promote market demand for biofuels and market incentives for their development and production; *regulations* like taxes or outright bans on existing products and services to open market spaces for new bio-based products and services; *procurement* rules to support new markets for bio-based products and services; *consumer labelling* to raise public and consumer awareness of bio-based products and services; new *standards and certification* schemes to identify relevant sustainability criteria for bio-based products and services; and *research funding and technology transfer* support to finance research and commercialization of new bio-technologies. It's notable that most countries pursue a range of MDPs at the same time (German Bioeconomy Council 2015a, b).

As the range and variety of MDPs across these different countries should indicate, there are many ways to support the emergence and expansion of new markets (e.g. biofuels) and the socio-technical transitions they are meant to engender (e.g. bio-economy). On a conceptual level, it's worth returning to Geels' (2002, 2005) conceptualization of societal transitions, as these MDPs highlight an ambiguity in the analytical framework. On the one hand, 'policy' and 'markets' are usually defined as parts

Table 6.1   Market development policies around the world

Policies	Details	Examples
Mandates	Covers biofuel blending mandates and renewable fuel standards (RFS) that require a particular percentage of biofuels in retail petroleum or specific volume of biofuel production	The USA has a RFS stipulating the production of 136 billion litres of biofuels by 2022; the EU's 2009 Renewable Energy Directive (RED) set a 10% target for biofuels in transport fuels by 2020
Subsidies	Covers range of subsidies for bio-based products and energy, including biofuels. These subsidies range from incentives for energy production through funding for demonstration plants to loans and grant support for facility construction	Germany and the UK provide financial support for the development of demonstration and pilot plants
Regulation	Covers regulations to ban products (e.g. plastic) as well as things like green taxes	The UK has a Climate Change Levy to tax commercial energy consumption; Italy introduced a 2011 law banning non-biodegradable plastic bags
Procurement	Covers a variety of schemes designed to encourage government purchasing of bio-based products and biofuels	In 2014 the USA reworked its BioPreferred programme to support the bio-economy through government purchasing
Research funding	Covers basic and applied research funding	Most countries have research support specifically directed at areas like biotechnology, biofuels, renewable energy, bio-based products, etc.; the EU, for example, has focused a significant proportion of FP7 funding on the cross-cutting theme of the 'knowledge-based bio-economy'
Technology transfer	Covers policies designed to encourage the commercialization of new research, often undertaken in public organizations (e.g. universities)	The UK established a Technology Strategy Boards in the of industrial biotechnology; the EU established a number of public-private partnerships including the Biobased Industries Consortium

(continued)

Table 6.1   (continued)

Policies	Details	Examples
Standards and certification	Covers the establishment of standards and certification for new products and services, especially where this might involve the incorporation of sustainability criteria	The EU has established standards for bio-based products (e.g. CEN/TC 411)
Labelling	Covers the creation of labels designed to create greater consumer awareness of new products and their sustainability characteristics	Germany established a certification scheme for renewable biological resources and 'Blue Angel' label for consumer products; France has established the label *batiment biosourcé* for bio-based buildings

Source: Germany Bioeconomy Council (2015a)

of the prevailing and dominant socio-technical regime; on the other hand, Geels argues that technological niches—where new socio-technical regimes emerge—require the resources "provided by public subsidies and private strategic investments" to lead to societal transformations (Geels 2005: 450). Similarly, Schot and Geels (2008: 537) note that "new technologies cannot immediately compete on the market against established technologies". In the context of the bio-economy and advanced biofuels then, MDPs are necessarily embedded within and yet conflict with the prevailing and dominant socio-technical regime (e.g. fossil fuels). As such, each socio-technical regime is assumed to have the seeds of its own destruction buried within it.

Another more important point to raise is that these MDPs illustrate the extent to which new markets represent an 'instituted process'—or institutional fixes for environmental problems—in the sense that they cannot be considered as neutral (or naturalized as some sort of pre-given social ontology) with consumer or user preferences and individual rational decision-making driving their emergence. This is important, as Schulz and Bailey (2014: 286) note, because there's a:

> ...current tendency for environmental social science and policy to focus on *individuals* as a major source of change [which] has stifled debate on how personal behaviours develop in association with, and are influenced by, wider social milieus and economic contexts.

To these social and economic contexts, we might add the *material* milieu of policies and markets. It's important, therefore, to examine how MDPs are framed and deployed in policy and business terms, as well as how they are configured by different materialities—that is, how environmental processes and economic processes shape each other, as emphasized by the environmental economic geography and other literatures (e.g. Bridge 2008, 2009; Mitchell 2011; Birch and Calvert 2015; Birch 2016a). In order to undertake this sort of analysis, I now examine the MDPs enacted to promote advanced biofuels in Canada, using this as a case study of the bio-economy with the aim of unpacking the policies shaping this new market and the materialities that limit the impacts of these MDPs.

Canada's Bio-Economy Strategy?

As discussed in Chap. 4, Canada doesn't have a distinct national bio-economy strategy, or even policy vision. This was a point made by a number of informants: for example, "We don't have, even in Canada we don't have a federal framework to support bioeconomy" (Provincial Ministry #110). Recent attempts to develop a more coherent and focused bio-economy agenda at the federal scale led to the establishment of a Bioeconomy Interdepartmental Working Group (BIWG) in the early 2010s. However, the BIWG has not produced a federal policy strategy at the time of writing in early 2018. While there may be no federal government policy vision, framework, or strategy in Canada, there are several examples of business and provincial government policy agendas. I've already mentioned some of the business-led agendas, such as the action plans and blueprints produced by BIOTECanada (2009) and CRFA (2014), but alongside these two examples, there are other business-led policy agendas in Canada. For example, in 2012 several trade associations in the automotive, biotechnology, bioenergy, chemical, and forest product sectors established the Bio-economy Network (BEN)—it is unclear, however, whether the BEN still functions today. Examples of provincial government strategies include BC Bioeconomy in British Columbia; BioEconomy Alberta; and Ontario BioAdvantage (e.g. BC Bioeconomy 2011). Obviously, each strategy has its own dynamic and focus; for example, Ontario's is more focused on agriculture, while British Columbia's is more focused on forestry. There is some coordination across the provinces through things like the Canadian Council of Forest Ministers, which recently produced a report called *A Forest Bioeconomy Framework for Canada* (CCFM 2017).

Despite the lack of a coherent national 'bio-economy' strategy, it's important to note that several federal agencies and ministries have implemented a number of policies designed specifically to promote and support the biofuels market, whether conventional or advanced. According to Luckert (2014), for example, there have been three waves of biofuels policy in Canada. First, in the 1990s, they centred on excise tax exemptions on ethanol (i.e. conventional biofuels); second, after 2003, they centred on subsidizing biofuels producers in order to create a domestic industry, although primarily focusing on ethanol producers; and finally, from the late 2000s, they have centred on establishing biofuels mandates (see Kedron 2015). I outline a number of these federal biofuels policies here and come back to more specific examples in the next section:

- *Ethanol Expansion Program*: established in 2003, this was an early attempt to support domestic biofuels production by financing the expansion of ethanol production capacity through a loan scheme (Laan et al. 2011).
- *Biofuels Opportunity for Producers Initiative* (BOPI): ran from 2004 until 2008 out of the Advancing Canadian Agriculture and Agri-Food (ACAAF) programme. It provided support for farmers to develop and/or increase biofuels production and represented an attempt to encourage the use of agricultural feedstock (e.g. wheat) in the production of (conventional) biofuels.
- *Renewable Fuel Regulations* (RFR): a biofuels mandate legislated in 2006 by the federal government and implemented in 2010 for petroleum/gasoline (5 percent renewable content) and in 2011 for diesel and heating distillate oil (2 percent). The RFR established a national renewable fuel mandate, although several provinces introduced earlier mandates; for example, Ontario introduced a 5 percent ethanol mandate through the 2005 *Ethanol in Gasoline* regulation implemented as part of a 2007 amendment to the *Environmental Protection* Act (1990) (Mondou and Skogstad 2012).
- *ecoAgriculture Biofuels Capital Initiative* (ecoABC): ran from 2007 until 2012, and was designed to support, through repayable contributions, the building or expansion of biofuels refineries.
- *NextGen Biofuels Fund*: part of Sustainable Development Technology Canada (SDTC) and running from 2007 until 2017. It was meant to support the commercialization of research by financing the construction of demonstration plants—at 40 percent of eligible costs—for

producing advanced biofuels and co-products, especially cellulosic ethanol. A similar scheme by SDTC called the *Sustainable Development Technology Fund* is designed to support commercialization of cleantech, which would include advanced biofuels.

- *ecoENERGY for Biofuels*: a programme running from 2008 until 2017. It was designed to provide an operating incentive to Canadian biofuels producers (C$0.10/litre of ethanol and C$0.20/litre for biodiesel). The policy intention was to foster a domestic Canadian biofuels industry, especially in advanced biofuels.
- *Forest Innovation Program*: introduced in 2012 as a way to support the forest industry diversify into new bio-product markets through the provision of financing for research and commercialization.

As should be evident from this range of policies, attempts to support biofuels cut across different government ministries, from agriculture through environment to research and innovation. As with other biofuels policies pursued in other countries, the main driver behind biofuels—and the wider bio-economy—has shifted over time, so that early concerns with finding outlets for agricultural surpluses shift towards environmental considerations before ending with economic development concerns. A similar trend in shifting priorities can be witnessed in the EU and USA.

MARKET DEVELOPMENT POLICIES IN ADVANCED BIOFUELS MARKET: CANADIAN CASE STUDY

In this final section of the chapter, I analyse the market development policies (MDPs) used to institute an advanced biofuel markets in Canada. I specifically focus on four key MDPs, namely: mandates; subsidies; feedstock supply chains; and standards and regulations. In analysing the way these MDPs are deployed in Canada, my aim is to examine how new markets are instituted and the biophysical materialities that limit this process.

Biofuel Mandates

Mandates refer to the introduction of renewable fuel standards or targets that stipulate content requirements for petroleum/gasoline. Some mandates like the USA's Renewable Fuel Standard (RFS) are volumetric mandates (e.g. billion gallons by a certain date), while others like the EU's Renewable Energy Directive (RED) are proportional blending ones. Canadian mandates tend to be proportional ones, whether federal or provincial. Such proportional mandates require a certain percentage of petroleum/gasoline sold to consist of biofuels, sometimes differentiating between conventional and advanced biofuels. In the Canadian context, the federal government's Renewable Fuel Regulations (RFR), enacted in 2006 and implemented in 2010/2011, is the main mandate, although provinces have their mandates as well. As Laan et al. (2011: 2) note, mandates "provide a guaranteed source of demand for biofuels, regardless of cost"—essentially, mandates help to create a market for a certain amount of biofuels. It's important to remember that most mandates are blending mandates and not regulations for the outright substitution of fossil fuels with biofuels. For example, petroleum/gasoline sold in Ontario is meant to contain 5 percent ethanol (either conventional or advanced)—that means 95 percent of the fuel sold is still derived from oil.

Some mandates contain regulations preferential to advanced biofuels, such as cellulosic biofuels; for example, some mandates let producers count advanced biofuels as double the equivalent of conventional biofuels. The development of markets for advanced biofuels is largely driven by these mandates, as one informant pointed out:

> I think most of the projects that we have seen in regards to biofuels platforms are probably due to the fact that people saw a trend towards a mandate coming into play, which would justify making a change in developing these platforms because there would be a market pull for it. (Federal Agency #106)

Moreover, another informant noted that "we [Canada] don't have a cellulosic ethanol mandate"—in contrast to countries like the USA— meaning that "the opportunity to develop cellulosic material for biofuel would be limited" (Civil Society Association #94). As these two informants indicate, mandates are an important mechanism for creating market demand as an incentive for biofuels producers. This is a point reiterated by this informant:

> I don't think that the advanced biofuels yet are far enough developed to be any kind of a competition [to conventional biofuels]. I think until there are actually policies to develop that market, like there have been for the first-generation ones, I think it'll be difficult for them to get off the ground. (Federal Agency #105)

In this sense, 'policies', especially mandates, are deeply implicated in the development of new socio-technical regimes through creating new markets, a point noted in the sustainability transitions literature (e.g. Geels 2005; Schot and Geels 2008). However, there is an important question here; will policy-makers, who are part of the prevailing socio-technical regime, promote the development of new regimes that threaten incumbents? One answer might be that wider socio-material conditions (e.g. climate change) will drive change in incumbent regimes (e.g. fossil fuels producers), implying that these wider social and material arrangements are a key influence on the development of new markets.

Taking these socio-material arrangements seriously necessitates an analysis of how they impact the introduction of mandates, as a MDP—that means analysing the interaction and inter-relation between environmental and economic processes (Bridge 2008). For example, one material consideration particular to the Canadian context is the biophysical effects of (extreme) cold on biofuels. An informant from a Provincial Ministry (#110) pointed out that the Ontario biofuel mandate excludes northern Ontario because of "the climate constraints for northern Ontario primarily that result from gelling [of biodiesel]". Consequently, mandates are not, by themselves at least, going to create an advanced biofuels market alone, since other, material factors impact these developments as well. As the same informant also pointed out, problems with gelling likely necessitate investment in new physical infrastructure—that is, a new material and spatial arrangement—in order to ensure that certain "products are useable" at all times of year (#110). As I argue elsewhere (e.g. Birch and Calvert 2015), this is illustrative of the ways that new technology markets are entangled with specific socio-material configurations.

Mandates are not enough to create markets; it's not enough for a government to simply stimulate demand for biofuels, although such policies play a significant role. Nevertheless, it's evident that the conditions for new markets are mediated through the state and policy action, meaning that markets can be considered as very much the product of politics and power (Tyfield 2014, 2017)—and perhaps not the replacement of the latter with the former as suggested by some writers on neoliberalism (e.g. Davies 2014). For example, one anecdote repeated to me more than once during my research was that the American ethanol industry has become so powerful because Iowa is the first primary in any presidential bid, meaning that all presidential hopefuls have to concern themselves with the politics of corn (see Gillon 2010). Here the geographical materialities of a feedstock—its cultivation in particular parts of a country—could potentially constrain the development of new markets, especially in advanced biofuels.

Subsidies

In Canada there are numerous subsidies—or subsidy supports—for conventional and advanced biofuels; see previous section for some examples. These subsidies range from more indirect support through excise tax exemptions and production credits to more direct support through grants, loans, and guarantees for specific projects or facilities. Canada phased out its tax exemption scheme in 2008 and brought in production credits as a way to support the growth of its domestic biofuels sector; the reason for this was because tax exemptions could be claimed by importers as easily as domestic producers, so did little to support domestic producers. Such subsidies represent another critical MDP, like mandates, and have underpinned the development of the conventional and advanced biofuels market in Canada. According to one informant, for example, there have been four phases to Canadian biofuel policy-making: first, regulations like mandates help establish market demand for conventional biofuels (see above); then capital incentive programmes help support the construction or upgrading of refining facilities; then production incentives support the expansion of conventional biofuels markets; and finally, novel technology funding supports the development of advanced biofuels (Federal Agency #105). Although rather linear, this narrative framing is illustrative of the general view of how to institute a market for advanced biofuels.

There is an obvious emphasis on production throughout these schemes. Production credits, for example, represent a subsidy per litre of biofuels produced and were introduced at the federal level in 2008 through the *ecoENERGY for Biofuels* programme, which ended in 2017 (Laan et al. 2011). Some provinces introduced their own production credit schemes. As with mandates, biofuels—especially advanced biofuels—are dependent on these sorts of MDP, as this informant argued:

> I don't think that's [cellulosic ethanol] a particularly viable technology if you strip away subsidies. One of the big problems is in that is pre-treatment, the cost of pre-treating the biomass before you start treating it. (Trade Association #101)

Other more direct, although project-specific, production subsidies include grants, loans, and guarantees introduced though policies like *Biofuel Opportunities for Producers Incentives* (BOPI) and *ecoAGRCIUL-TURE Biofuel Capital* (ecoABC) (ibid.). These sorts of scheme supported the building of new refineries through grants, loans, business assistance, and research, cutting across the value chain and its infrastructure requirements (see next chapter).

As noted already, subsidies represent another MDP, but one that is predominantly focused on the production side of biofuels markets: for example, tax exemptions, production incentives, capital grants and loans, depreciation, and so on (Laan et al. 2011: 69). Canada does not necessarily have a range of consumption-based subsidies or incentives, meaning that it is very much driven by the demands of producers in the value chain. As a result, the deployment of subsidies as MDPs has led to some problems, as one informant pointed out:

> I think our policy suite in [Province], specifically, is kind of fragmented right now and really not working on all aspects of the value chain, looking at all products and, you know, indicating different sectors into the biobased sectors ... other jurisdictions have a lead in that, especially the European Union. (Provincial Ministry #110)

Other subsidies at the consumption end of the value chain, like excise tax exemptions or carbon taxes, have very different implications. But as another informant noted, a carbon tax like the one introduced in British Columbia "doesn't necessarily incent the bio-component versus the non-renewable component" (Consultancy #93).

As an issue I've already raised in the last chapter, the production focus of current subsidies ignores the need to support biofuels consumption and the likely dramatic socio-material transformation this will require. It'll either necessitate the restructuring and reconfiguration of existing infrastructures (e.g. processing, blending, pipelines, pumps, etc.) at huge social costs, or the search for ways to substitute biofuels for fossil fuels within existing infrastructure (see Chap. 5). The former represents a risky strategy for policy-makers, as an informant highlighted:

> That model [building a whole new system], by the way, was tried in Germany about 10 years ago where they put such financial incentive into the market for using 100% biodiesel that the market created a parallel infrastructure. So they had separate tanking, distribution trucks and pumps ... then the government

> sort of came to the realization that the subsidy by taking taxes off of diesel, the inherent subsidy was unsustainable and they ratcheted back and that whole parallel infrastructure system was collapsed. (Consultancy #93)

The latter represents the pursuit of technological pathways like 'drop-in' biofuels, which don't radically alter the dominant socio-technical regime (Birch and Calvert 2015). However, drop-in biofuels aren't likely to alter other (problematic) aspects of prevailing energy systems like suburban lifestyles and automobile dependence, modern supply chain systems, heavy trucking, and so on.

Here, there's a major tension in how biofuels are legitimated, which I discussed in the last chapter, but which it's worth considering again now. Any attempt to replace the prevailing energy market with biofuels—advanced or otherwise—won't necessarily reduce overall GHG emissions without significantly raising infrastructure costs or expanding the bio-economy as a societal transformation. As one informant noted:

> … if the bio-economy is about moving through a biological system from a fossil fuel-based system, then we have to talk at the scale of a fossil fuel-based system … It's [bio-economy] too small. And part of it is this understanding of the scale of our fossil fuel energy system. If we're going to have a bio-economy, we've got to be thinking in scale. (University #96)

This is especially the case when it comes to advanced biofuels, since the economic feasibility of individual facilities is too volatile because the scale of biomass usage is so small and its supply chain configuration (e.g. trucking) is dependent on fossil fuels whose pricing ends up driving biofuels pricing (Calvert et al. 2017). This point was made by one informant:

> And part of the challenge obviously is that biomass in a lot of markets is a very small component of actual supply, so transportation fuels, electricity, woody, you know, biomass would constitute a very small proportion of the market. And therefore, the pricing of that market is not dictated by biomass, it's dictated by something else. And so the pricing is, the [biofuel] producers have no control over what's going on. (University #90)

Feedstock Supply Chains

Market uncertainty is often cited as a reason why the advanced biofuels market hasn't developed as quickly as expected or hoped. As noted elsewhere in this book, advanced biofuels are still largely a socio-material system in development, they aren't a viable alternative or substitute for fossil fuels. A key reason for this lack of viability relates to the supply of different feedstocks—that is, biomass like crops, forest products, residues, and organic waste. Feedstock supply chains represent the organization and configuration of access to and availability of this biomass for biofuels producers, whether they are conventional or advanced biofuels. Producers are reliant on securing consistent and long-term supplies of biomass, of whatever variety; as such, there's no use building a biofuels refinery if there's no biomass supply for that refinery (Calvert et al. 2017). As several informants stressed during interviews with me, biofuels producers are dependent on the socio-material configuration of feedstock supply in order to operate. As such, MDPs need to ensure secure feedstock supply through economic (e.g. contracts) and technological (e.g. 'feedstock agnostic' conversion processes) means.

First, this means access to biomass nearby, since the cost of feedstock increases as it's transported over longer and longer distances. Second, this entails "loyalty of supply", or, as one informant put it, "if you don't have a market for your biomass, you're not going to produce it" (Civil Society Association #94). Both biomass suppliers and biofuels producers are materially dependent on one another, in ways that might militate against market-like exchange; rather, they both need long-term contractual arrangements in order to ensure that each of them can operate. Third, and according to two informants (Federal Agency #106 and Trade Association #107), the wider bio-economy is going to be increasingly dependent on the conversion of diverse feedstocks (e.g. crops, woods, grass, residues, waster) into a fungible commodity (i.e. sugars) that can be used as an input into a range of processes (e.g. biofuels production, biochemical production, etc.) (Shortall et al. 2015). I come back to this in the next chapter.

Feedstock supply has particular materialities. Different feedstock has very different biophysical characteristics, meaning that harvesting, processing, and transportation are configured by things like seasonal growing speeds, water content, energy density, and so on (Mabee 2007). As one informant put it:

> ...there's a large variety of feedstocks, right? So in the forest you've got round wood, you've got harvest residues. Then you get to the mill and you've got a variety of mill residues. You've got hog fuel. You've got chips, which are often used for pulp but not always. And you've got sawdust and you've got shavings. And each of those has got different properties and each of those has got different values to different people. (University #90)

As an example, feedstock materialities entail certain implications for feedstock transport and the infrastructure needed to get feedstock from harvest or collection points to processing. Leaving aside the role of pre-processing (e.g. pelleting) here (see Birch and Calvert 2015), the material-ity of feedstock goes beyond their biophysical form (e.g. size of material). Certain feedstocks are already integrated into existing infrastructure sys-tems, meaning that there's no requirement to create new material and spatial arrangements, while others are not. For example, municipal solid waste (MSW) feedstock is already integrated into a collection infrastructure (e.g. garbage collection), reducing the cost of collection in that particular case. Other feedstock (e.g. timber residues), in contrast, require the intro-duction of new infrastructure and logistics, or their retrofitting, adding to the cost of developing that specific advanced biofuels markets (e.g. cellulosic).

All of this suggests that the neoliberalization of nature—as discussed in Chap. 2 and earlier in this chapter—needs to be nuanced so that it's not characterized as the simple installation of markets in the organization of nature-economy relations; when it comes to biofuels at least, there are significant limits to the role that market exchange *can* play in the roll-out of biofuels as a societal transition, especially advanced biofuels. 'Too much' market is likely to disrupt this transition pathway because it disrupts the necessary and long-term 'lock-in' needed between feedstock supply and biofuels production.

Standards and Regulations

A final set of MDPs worth considering are standards and regulations, especially those that are distinct from mandates. Many MDPs are shaped by the interests of producers, as I note above with subsidies. When it comes to consumers, MDPs are generally associated with the idea of pro-moting "public acceptance" (Consultancy #93) or "market readiness" (Provincial Ministry #110). Standards and regulations are characterized as key to this acceptance or readiness. Generally speaking, they are an impor-tant component of any socio-technical regime, reflecting the rules that help different social actors coordinate their activities (Geels 2005). For example, Daemmrich (2015: 50) discusses a relevant case where "the writ-ing of new technical standards" was "an integral part of introducing new biodegradable polymers to the market". According to Daemmrich, the standards proactively developed alongside a new bio-product and thereby created a market for that product. When it comes to transportation fuels, such standards are necessarily tied to the requirements of the combustion engine, which has material implications for understanding the emergence of biofuels markets.

At present, advanced biofuels don't have "any kind of universally accepted standards" according to one informant (Federal Agency #105), especially when it comes to the development of standards at the International Organisation for Standardization (ISO). For example, *ISO/ TC 28/SC 7 Liquid Biofuels* is still in development, although the ISO committee was established in 2007. As this informant commented:

> In 2009 both my colleagues are involved in the ISO sustainability of bioenergy standard development, it's in years four and five already. It's theoretically going to wrap up this year or next. (Consultancy #93)

Part of the reason for the delay in developing this standard is because it's (politically) difficult to integrate sustainability criteria into the standard; however, according to many policy-makers, such criteria are necessary for constructing markets in which consumers can make choices about what products to buy based on their individual environmental preferences (Daemmrich 2015; German Bioeconomy Council 2015a). The issue with developing such sustainability standards is their relationship to the materiality of biofuels. The informant above went on to note that incorporating sustainability into such standards is important for creating "public acceptance" because it enables consumers to compare different biofuels (e.g. palm oil versus corn versus cellulosic) (#93). However, all biofuels have to meet the similar biophysical quality standards (so they can be used in all combustion engines), but sustainability criteria—and the markets they aim to institute—necessarily go beyond the biophysical qualities of a product alone.

Advanced biofuels are designed for use in prevailing transportation systems, including combustion engines as well as the back-end infrastructure (e.g. storage, refining, blending, shipping, etc.). As such, the standards to which they are held predominantly relate to quality criteria, *standardizing* for use in combustion engines rather than for sustainability. The critical point to make here is that standards and the criteria on which those standards are measured relate to the biophysical qualities of the fuel itself (e.g. gross heat combustion, water content, density, sulphur content, etc.); for example, see the *ASTM D7544 – 12 Standard Specification for Pyrolysis Liquid Biofuel*. As such, these biophysical standards are embodied in the product itself. In contrast, according to one informant, the development of sustainability standards for biofuels involves a broader set of accounting issues relating to the feedstock (e.g. timeframe of replanting), social conditions of harvesting (e.g. labour standards), environmental performance (e.g. life cycle GHG emissions), and so on (University #96). As a result, the development of sustainability standards for biofuels involves a series of arguments about the development, measurement, and monitoring of a range of social, political, and economic processes, not just embodied ones (see Chap. 7). However, just considering the embodied criteria alone would not facilitate the creation of an advanced biofuels market premised on the notion that they offer better sustainability outcomes than petroleum or conventional ethanol. Yet, the sustainability standards cannot be embodied in the materiality of the biofuels itself, as they are inherently social, political, and economic judgements.

CONCLUSION

Many countries are developing bio-economy strategies to promote a transition from fossil fuel dependent economies to bio-based economies. An important component in these strategies is the development of advanced biofuels as a (partial) replacement for petroleum. So far, debates on these types of societal or sustainability transition emphasize changes to socio-technical regimes resulting from new technological niches (e.g. Geels 2002). Less attention has been given to the material and spatial configuration represented by the socio-material configuration of new markets, meaning that the implications of materiality for transitions have been relatively under-theorized and under-studied. It's important, however, to integrate materiality conceptually into the analysis of socio-technical systems and their transitions in order to understand how biophysical processes, qualities, and characteristics impact on social, political, and economic processes. In order to undertake this analysis, I examined market development policies (MDP) designed to promote advanced biofuels in Canada in this chapter.

My theoretical aim was to conceptualize the materialities of societal or sustainability transitions in order to extend the analytical lens to the biophysical qualities, characteristics, and processes of different socio-technical regimes (Geels 2002, 2004, 2005; Schot and Geels 2008). While this materiality is currently integrated into sustainability transitions as part of the 'landscape' (e.g. Geels 2005), this misses the point that *all* components of socio-technical regimes *and* technological niches are necessarily material. Market/user preferences are shaped by biophysical processes that limit the emergence of new socio-technical regimes; for example, electric cars have physical limits that restrict their popularity, not simply because of their technological qualities (e.g. battery life) but also because they are only one component in a complex social and material environment (e.g. suburban life, automobile commuting, flexible employment, etc.) (Tyfield 2014). Consequently, it's important to theorize sustainability transitions as *socio-material systems* by more fully integrating biophysical processes into the analysis (Birch 2013, 2016a).

My initial focus in the chapter was on the development of bio-economy strategies around the world and the deployment of specific MDPs in different countries as ways to develop new bio-based markets. In order to explore nature-economy relations in more depth, however, I presented a case study of the development of advanced biofuels in Canada as a way to examine how environmental and economic processes interact. In discussing the specific MDPs implemented in Canada as part of the policy attempt to create an advanced biofuels market, I sought to analyse the implications of the specific biophysical materialities of advanced biofuels to the construction and emergence of new markets. My intention was not to suggest that advanced biofuels will never develop, in Canada or elsewhere, but to highlight the difficulties that face policy-makers and others seeking to kick-start new transition pathways. It's important, in any context, to consider the materialities of transitions, or they are likely to derail attempts to stimulate or create new economic sectors.

References

Aguilar, A., Wohlgemuth, R. and Twardowski, T. (2018) Preface to the special issue bioeconomy, *New Biotechnology* 40: 1–4.

Bakker, K. (2009) Commentary: Neoliberal nature, ecological fixes, and the pitfalls of comparative research, *Environment and Planning A* 41: 1781–1787.

Bakker, K. (2010) The limits of 'neoliberal natures': Debating green neoliberalism, *Progress in Human Geography* 34(6): 715–735.

Bauer, F. (2018) Narratives of biorefinery innovation for the bioeconomy – Conflict, consensus, or confusion?, *Environmental Innovation and Societal Transitions*, doi.org/10.1016/j.eist.2018.01.005.

BC Bio-economy (2011) *BC Bio-economy*, Victoria: British Columbia Provincial Government, retrieved from: http://www.gov.bc.ca/jtst/down/bio_economy_report_final.pdf.

BIOTECanada. (2009) *The Canadian blueprint: Beyond moose & mountains. How we can build the world's leading bio-based economy*, Ottawa: BioteCanada.

Birch, K. (2013) The political economy of technoscience: An emerging research agenda, *Spontaneous Generations: A Journal for the History and Philosophy of Science* 7(1): 49–61.

Birch, K. (2016a) Materiality and sustainability transitions: Integrating climate change into transport infrastructure in Ontario, Canada, *Prometheus: Critical Studies in Innovation* 34(3–4): 191–206.

Birch, K. (2016b) Emergent policy imaginaries and fragmented policy frameworks in the Canadian bio-economy, *Sustainability* 8(10): 1–16.

Birch, K. and Calvert, K. (2015) Rethinking 'drop-in' biofuels: On the political materialities of bioenergy, *Science and Technology Studies* 28(1): 52–72.

Birch, K. (2017) Introduction: Techno-economic assumptions, *Science as Culture* 26(4): 433–444.

Birch, K., Levidow, L. and Papaioannou, T. (2010) *Sustainable capital?* The neo-liberalization of nature and knowledge in the European knowledge-based bio-economy, *Sustainability* 2(9): 2898–2918.

Birch, K., Levidow, L. and Papaioannou, T. (2014) Self-fulfilling prophecies of the European knowledge-based bio-economy: The discursive shaping of institutional and policy frameworks in the bio-pharmaceuticals sector, *Journal of the Knowledge Economy* 5(1): 1–18.

Birch, K., Peacock, M., Wellen, R., Shenaz Hossein, C., Scott, S. and Salazar, A. (2017) *Business and society: A critical introduction*, London: Zed Books.

Bridge, G. (2008) Environmental economic geography: A sympathetic critique, *Geoforum* 39, 76–81.

Bridge, G. (2009) Material worlds: Natural resources, resource geography and the material economy, *Geography Compass* 3(3), 1217–1244.

Brown, W. (2015) *Undoing the demos*, Cambridge, MA: Zone Books.

Bugge, M., Hansen, T. and Klitkou, A. (2016) What is the bioeconomy? A review of the literature, *Sustainability* 8: 1–22.

Calvert, K., Birch, K. and Mabee, W. (2017) New perspectives on an old energy resource: Biomass and emerging bio-economies, in S. Bouzarovski, M. Pasqualetti and V. Castan Broto (eds) *The Routledge research companion to energy geographies*, London: Routledge, pp. 47–60.

Castree, N. (2008a) Neoliberalising nature: The logics of deregulation and reregulation, *Environment and Planning A* 40: 131–152.

Castree, N. (2008b) Neoliberalising nature: Processes, effects, and evaluations, *Environment and Planning A* 40: 153–173.

CCFM. (2017) *A forest bioeconomy framework for Canada*, Ottawa: Canadian Council of Forest Ministries.

CEC. (2005) *New perspectives on the knowledge-based bio-economy: Conference report*, Brussels: DG-Research, Commission of the European Communities.

CEC. (2012) *Innovating for sustainable growth: A bioeconomy for Europe* {SWD(2012) 11 final}, Brussels: European Commission.

Chisti, Y. (2010) Editorial: A bioeconomy vision of sustainability, *Biofuels, Bioproducts & Biorefining* 4(4): 359–361.

CRFA. (2014) *Evolution and growth: From biofuels to bioeconomy*, Ottawa: Canadian Renewable Fuels Association.

Coenen, L., Benneworth, P. and Truffer, B. (2012) Towards a spatial perspective on sustainability transitions, *Research Policy* 41(6): 968–979.

Collard, R-C., Dempsey, J. and Rowe, J. (2016) Re-regulating socioecologies under neoliberalism, in S. Springer, K. Birch and J. Macleavy (eds) *The handbook of neoliberalism*, New York: Routledge, pp. 455–465.

Daemmrich, A. (2015) Anticipatory markets: Technical standards as a governance tool in the development of biodegradable plastics, in S. Borras and J. Edler (eds) *The governance of socio-technical systems*, Cheltenham: Edward Elgar, pp. 49–69.

Davies, W. (2014) *The limits of neoliberalism*, London: Sage.

de Besi, M. and McCormick, K. (2015) Towards a bioeconomy in Europe: National, regional and industrial strategies, *Sustainability* 7: 10461–10478.

El-Chickakli, B., von Braun, J., Lang, C., Barben, D. and Philp, J. (2016) Five cornerstones of a global bioeconomy, *Nature* 535: 221–223.

Fargione, J., Hill, J., Tilman, D., Polasky, S. and Hawthorne, P. (2008) Land clearing and the biofuel carbon debt, *Science* 319: 1235–1238.

Frow, E., Ingram, D., Powell, W., Steer, D., Vogel, J. and Yearley, S. (2009) The politics of plants, *Food Security* 1(1): 17–23.

Geels, F. (2002) Technological transitions as evolutionary reconfiguration processes: A multi-level perspective and case study, *Research Policy* 31: 1257–1274.

Geels, F. (2004) Understanding system innovations: A critical literature review and a conceptual synthesis', in B. Elzen, F. Geels and K. Green (eds.) *System innovation and the transition to sustainability: Theory, evidence and policy*, Cheltenham: Edward Elgar, pp. 19–47.

Geels, F. (2005) The dynamics of transitions in socio-technical systems: A multi-level analysis of the transition pathway from horse-drawn carriages to automobiles (1860–1930), *Technology Analysis and Strategic Management* 17(4): 445–476.

Geels, F. and Schot, J. (2007) Typology of sociotechnical transition pathways, *Research Policy* 36(3): 399–417.

German Bioeconomy Council. (2015a) *Bioeconomy policy: Synopsis and analysis of strategies in the G7*, Berlin: Office of the Bioeconomy Council.

German Bioeconomy Council. (2015b) *Bioeconomy policy (Part II): Synopsis of national strategies around the world*, Berlin: Office of the Bioeconomy Council.

Gillon, S. (2010) Fields of dreams: Negotiating an ethanol agenda in the Midwest United States, *The Journal of Peasant Studies* 37: 723–748.

Golembiewski, B., Sick, N. and Broring S. (2015) The emerging research landscape on bioeconomy: What has been done so far and what is essential from a
technology and innovation management perspective, *Innovative Food Science
and Emerging Technologies* 29: 308–317.

Hansen, T. and Coenen, L. (2015) The geography of sustainability transitions:
Review, synthesis and reflections on an emergent research field, *Environmental
Innovation and Societal Transitions* 17: 92–109.

Hardin, G. (1968) The tragedy of the commons, *Science* 162: 1243–1248.

Hausknost, D., Schriefl, E., Lauk, C. and Kalt, G. (2017) A transition to which
bioeconomy? An exploration of diverging techno-political choices, *Sustainability*
9(4): 1–22.

Hudson, R. (2008) Cultural political economy meets global production networks:
A productive meeting? *Journal of Economic Geography* 8(3): 421–440.

Kedron, P. (2015) Environmental governance and shifts in Canadian biofuel production and innovation, *The Professional Geography* 67(3): 385–395.

Laan, T., Litman, T. and Steenblik, R. (2011) *Biofuels–At what cost? Government
support for ethanol and biodiesel in Canada* [updated version], Geneva:
International Institute for Sustainable Development.

Levidow, L., Papaioannou, T. and Birch, K. (2012a) Neoliberalising technoscience
and environment: EU policy for competitive, sustainable biofuels, in
L. Pellizzoni and M. Ylönen (eds) *Neoliberalism and technoscience: Critical
assessments*, Farnham: Ashgate Publishers, pp. 159–186.

Levidow, L., Birch, K. and Papaioannou, T. (2012b) EU agri-innovation policy:
Two contending visions of the knowledge-based bio-economy, *Critical Policy
Studies* 6(1): 40–66.

Levidow, L., Birch, K. and Papaioannou, T. (2013) Divergent paradigms of
European agro-food innovation: The knowledge-based bio-economy (KBBE)
as an R&D agenda, *Science, Technology, and Human Values* 38: 94–125.

Loftus, A. and Budds, J. (2016) Neoliberalizing water, in S. Springer, K. Birch and
J. McLeavy (eds) *The handbook of neoliberalism*, New York: Routledge,
pp. 503–513.

Lohmann, L. (2016) Neoliberalism's climate, in Springer, S., Birch, K. and
MacLeavy, J. (eds) *The handbook of neoliberalism*, London, Routledge,
pp. 480–492.

Luckert, M.K. (2014) Market and government failures in the competition for land
for biofuel production in Canada, *Biofuels* 5(3): 211–218.

Mabee, W. (2007) Policy options to support biofuel production, *Advances in
Biochemical Engineering* 108: 329–357.

Mabee, W., Gregg, D. and Saddler, J. (2005) Assessing the emerging biorefinery
sector in Canada, *Applied Biochemistry and Biotechnology* 121: 765–778.

Mathews, J.A. (2009) From the petroeconomy to the bioeconomy: Integrating
bioenergy production with agricultural demands, *Biofuels, Bioproducts &
Biorefining* 3: 613–632.

McCormick, K. (2014) The bioeconomy and beyond: Visions and strategies, *Biofuels* 5(3): 191–193.

McCormick, K. and Kautto, N. (2013) The bioeconomy in Europe: An overview, *Sustainability* 5: 2589–2608.

Mirowski, P. (2013) *Never let a serious crisis go to waste*, Cambridge MA: Harvard University Press.

Mitchell, T. (2011) *Carbon democracy*, London: Verso.

Mondou, M. and Skogstad, G. (2012) *The regulation of biofuels in the United States, European Union and Canada*, University of Toronto: CAIRN.

Mukhtarov, F., Gerlak, A. and Pierce, R. (2017) Away from fossil-fuels and toward a bioeconomy: Knowledge versatility for public policy?, *Environment and Planning C* 35(6): 1010–1028.

OECD (2006) *The bioeconomy to 2030: Designing a policy agenda*, Paris: Organisation for Economic Co-operation and Development.

Patermann, C. and Aguilar, A. (2018) The origins of the bioeconomy in the European Union, *New Biotechnology* 40(A): 20–24.

Peck, J. (2010) *Constructions of neoliberal reason*, Oxford: Oxford University Press.

Pellerin, W. and Taylor, D.W. (2008) Measuring the biobased economy: A Canadian perspective, *Industrial Biotechnology* 4(4): 363–366.

Pfau, S.F., Hagens, J.E., Dankbaar, B. and Smits, A.J.M. (2014) Visions of sustainability in bioeconomy research, *Sustainability* 6: 1222–1249.

Polanyi, K. (1957) The economy as instituted process, in K. Polanyi, C. Arensberg and H. Pearson (eds) *Trade and market in the early empires*, Illinois: Free Press and Falcon's Wing Press, pp. 239–270.

Puddister, D. et al. (2011) Opportunities and challenges for Ontario's forest bioeconomy, *The Forestry Chronicle* 87(4): 468–477.

Richardson, B. (2012) From a fossil-fuel to a biobased economy: The politics of industrial biotechnology, *Environment and Planning C* 30(2): 282–296.

Scarlat, N. and Dallemand, J-F. (2011) Recent developments of biofuels/bioenergy sustainability certification: A global overview, *Energy Policy* 39: 1630–1646.

Scarlat, N., Dallemand, J., Monforti-Ferrario, F. and Nita, V. (2015) The role of biomass and bioenergy in a future bioeconomy: Policies and facts, *Environmental Development* 15: 3–34

Schulz, C. and Bailey, I. (2014) The green economy and post-growth regimes: Opportunities and challenges for economic geography, *Geografiska Annaler B* 96(3): 277–291.

Schmid, O., Padel, S. and Levidow, L. (2012) The bio-economy concept and knowledge base in a public goods and farmer perspective, *Bio-based and Applied Economics* 1: 47–63.

Schot, J. and Geels, F. (2008) Strategic niche management and sustainable innovation journeys: Theory, findings, research agenda, and policy, *Technology Analysis and Strategic Management* 20(5): 537–554.

Searchinger, T., Heimlich, R., Houghton, R.A., Dong, F., Elobeid, A., Fabiosa, J., Tokgoz, S., Hayes, D. and Yu, T.-H. (2008) Use of US croplands for biofuels increases greenhouse gases through emissions from land-use change, *Science* 319: 1238–1240.

Shortall, O., Raman, S. and Millar, K. (2015) Are plants the new oil? Responsible innovation, biorefining and multipurpose agriculture, *Energy Policy* 86: 360–368.

Sims, R., Mabee, W., Saddler, J. and Taylor, M. (2010) An overview of second generation biofuel technologies, *Bioresource Technology* 101: 1570–1580.

Sorda, G., Banse, M. and Kemfert, C. (2010) An overview of biofuel policies across the world, *Energy Policy* 38: 6977–6988.

Staffas, L., Gustavsson, M. and McCormick, K. (2013) Strategies and policies for the bioeconomy and bio-based economy: An analysis of official national approaches, *Sustainability* 5: 2751–2769.

TNI. (2015) *The bioeconomy: A primer*, Amsterdam: Transnational Institute.

Tyfield, D. (2014) Putting the power in 'socio-technical regimes' – E-mobility transition in China as political process, *Mobilities* 9(4): 585–603.

Tyfield, D. (2017) *Liberalism 2.0 and the rise of China*, London: Routledge.

White House (2012) *National bioeconomy blueprint*, Washington: The White House.

CHAPTER 7

Co-Constructing Markets and Natures in Bio-Economies

INTRODUCTION

When I first started researching the bio-economy—which was over a decade ago now (e.g. Birch 2006a)—my main analytical approach was to frame the bio-economy as a powerful policy or societal vision, narrative, or discourse that configured policy in particular ways. It could be both future oriented and backward looking (e.g. Birch 2006b), meaning that the bio-economy is bound up with all sorts of different expectations (e.g. Borup et al. 2006), anticipations (e.g. Hilgartner 2007), and assumptions (e.g. Birch 2017a). More recently, however, I've become interested in trying to unpack the implications of the biophysical materialities—of the underlying biological material—underpinning the bio-economy for our economies, societies, and polities. Driven by a conceptual approach drawing on environmental economic geography—see Chap. 2—I've been particularly concerned with the question of how to understand—and act on, in a policy sense—the constitutive inter-relationship between economic and environmental processes.

In this book, and the research on which it's based, this concern of mine has translated into the analysis of the development of advanced biofuels in Canada, as a representative case study of the emergence of the broader bio-economy. Advanced biofuels are a key technological solution—or 'techno-fix'—favoured by a range of stakeholders because they seemingly represent a 'win-win-win' outcome; they are meant to reduce greenhouse gas (GHG) emissions at the same time as they are meant to ensure energy security, *and* promote economic development, especially in peripheral,

159

K. Birch, *Neoliberal Bio-Economies?*,
https://doi.org/10.1007/978-3-319-91424-4_7

rural regions (Frow et al. 2009; Richardson 2012). As part of a socio-technical transition to a low-carbon future, advanced biofuels are meant to solve some pretty intractable environmental problems—although it is important to note that their own development seems intractable for socio-economic reasons.

Our societies need to reduce energy use, and to do so substantially right now. If we take the 'scary maths' presented by Bill McKibben (2012) as accurate, then the real cost of fossil energy—taking into account its environmental externalities—needs to rise significantly in a short space of time. As an estimate, fossil fuel energy prices need to rise fivefold in the next few years if everyone in the world wants to limit rising global temperatures to 1.5 °C, which was agreed upon at the 2015 Paris United Nations Climate Change Conference (or COP21). A fivefold price increase would, however, have significant impacts on a slew of social activities, behaviours, preferences, and choices. For example, in Canada, the GHG emissions from freight are expected to rise by 24 percent between 2005 and 2020 as Canada's society and economy become more dependent on the long-distant transport of goods via heavy-duty trucks (Wood and Mabee 2016). To reduce these sorts of environmental impacts will necessitate the increasing deployment of biofuels, especially advanced biofuels which produce fewer GHG emissions (ibid.).

In the past two chapters, I've analysed the (political-economic) materialities of biological matter—or biomass—in order to understand how these materialities both enable *and* limit the development of and the form taken by advanced biofuels as part of a socio-technical transition. On the one hand, Chap. 5 was my attempt to show how materialities enable and legitimate the development of a particular kind and form of biofuel, specifically 'drop-in' biofuels; for example, the cost of replacing existing and dominant energy infrastructures (e.g. pipelines, refineries, retail outlets, etc.) drives this biofuels development. On the other hand, in Chap. 6 my main argument was that the biophysical qualities of biofuels limit the construction, or instituting, of markets through the deployment of market development policies (MDPs); for example, economic uncertainties around feedstock supply necessitate long-term contractual arrangements that militate against (neoliberal) market arrangements. As these two chapters illustrate, in both cases, these materialities are as much political-economic (e.g. economic uncertainty, cost considerations) as they are biophysical (e.g. feedstock variability, physical infrastructure), suggesting that both markets and natures—in this book's parlance—are substantially and significantly correlated.

Rather than claim that the political-economic or the material side of these political-economic materialities is more important than the other (e.g. neoliberal processes shape the environment, or nature resists said neoliberal processes), my aim in this chapter is to analyse how the interaction between the two sides constitutes something new *and*, as important, something different from either the conception and representation of 'natures' or 'markets'. Analytically speaking, the interaction of economic and environmental processes does not entail the interaction of two distinct forces with one another; instead, my theoretical goal in this chapter is to show how markets and natures are co-constructed. As such, it's crucial to understand how 'nature' and 'market' are understood by social actors, but also to then think through how natures and markets are co-constructed rather than simply interacting forces. In order to do this, I start the chapter with a theoretical discussion of material markets and social natures drawing on scholarship in geography and other fields. I then discuss the co-construction of markets and natures in the development of advanced biofuels in Canada, before concluding with some wider implications for the bio-economy.

Social Natures, Material Markets

Instead of theorizing markets as necessarily problematic, my intention in this chapter is to think through how they result from nature-economy relations and from the very sociality and materiality that these relations entail. This means analysing how markets—often represented as primarily or purely social institutions—are co-constructed with a range of material knowledges, practices, and entities—these would include definitions of waste and sustainability, experiences of location and distance, and the generation of new landscapes and 'things'. Moreover, it means resisting the urge to theorize markets and materialities as in conflict; instead, my intent is to think through concepts like 'socio-natures' and 'material markets' when it comes to finding solutions to environmental problems.

Socio-natures

It's now commonly accepted across the political easel—at least in academic and policy circles—that 'nature' is not a blank canvas on which we can simply paint human activity, either in reality (for want of a better word) or in our analysis of said reality. Even neoclassical economists are now attempting to understand how human action is embedded within a broader ecological system (e.g. the World) (e.g. Stern 2010). A range of more heterodox thinkers arrived at this point some time ago, including geographers and political ecologists, who I come back to shortly. Basically, there is agreement that economic activities take place within—rather than alongside—a broader ecological system. However, it's also notable that most people now accept—with some obvious outliers—that humans also impact their environment, as much as the environment shapes human activities. Policy action on climate change is the clearest example of this acknowledgement, even though a few people still contest its anthropogenic basis. My focus here is on the work of political ecologists and geographers, leaving aside the work of others (e.g. ecological economists).

A particularly important point made by political ecologists—who are interested in the interaction between ecology and political economy—is that us humans have tended in recent history to ignore our own "dependence on nature" due to "the expansionist logic of a capitalist system" (Foster 2002: 9). Furthermore, this attitude highlights a major contradiction in capitalism itself; namely, it's not possible for capitalism to continually expand when there are very definite and finite environmental limits (e.g. Earth's ecological carrying capacity). As Jason W. Moore (2015) notes, it's possible to conceptualize capitalism as a 'world-ecology' in which capital accumulation is bound up with particular forms of ecological materialities, like cheap food, that come to shape the planet's environment—and in problematic ways, as evidenced by climate change. As with those approaching these issues from the 'neoliberal natures' perspective (see Chap. 2), political ecologists like John Bellamy Foster (2002: 32) tend to frame the market—and capitalism more generally—as an almost 'artificial' thing, contrary to natural, ecological processes. As such, markets are often presented as a direct cause of environmental problems, or an unresolvable contradiction.

Geographers have generally laid more stress on social-nature hybrids, or 'socio-natures', in their discussions. Some argue that 'nature' is a social construction, reflecting a series of visions, images, discourses, and similar (Robbins et al. 2010). As such, it isn't possible to identify 'nature' out there, outside of our collective agreement. Other geographers, especially those who draw on the Marxist political economy tradition, frame the discussion in terms of the 'production of nature'. For example, the geographer Neil Smith (1984) argued that it's important to think about nature as the product of political-economic processes like commodification, enclosure, and so on (also Heynen et al. 2006). Generally speaking, much of the geography literature is now defined by the idea that social and natural phenomena and processes are necessarily hybrid—that is, both material *and* political-economic (Bakker and Bridge 2006). It's not helpful, from this perspective, to refer to the 'social' or the 'natural' alone since, by definition, most phenomena and processes are 'socionatural' (Gellert 2005). This is a useful approach to take because it eschews the notion that humans produce nature, but similarly avoids the assumption that nature produces humans and their societies.

Material Markets

Throughout the literatures I'm looking at here, there is a tendency to treat the 'market' as an abstract mechanism, almost untouched by social or natural influences—and that applies to critics of neoliberalism and markets as much as neoclassical economists. In my view (see Birch 2017b), the very concept of 'neoliberalism' is fundamentally about the meaning and importance of markets in society, whether it's being used by critics or supporters. Frequently, critics and supporters of neoliberalism have the same core assumption; namely, (abstract) markets can be constructed and imposed or installed as the main organizing principle and institution in all areas of social life, including the management of environmental problems. As I've already noted in this book, such assumptions are premised on the notion that there are or can be markets that sit outside of all sorts of social, political, and ecological factors that shape—materially, culturally, ethically, or otherwise—the development and deployment of specific institutional arrangements, like markets. For example, influential geographers like Tickell and Peck (2003: 166) have defined neoliberalism as a process representing the "mobilization of state power in the contradictory extension and reproduction of market(-like) rule". While almost every social actor—be it politician, business person, NGO worker, environmentalist, citizen, and so on—is thereby implicated in the extension of market-like rule from their perspective, the foundational 'market' isn't polluted by such hybridity; the market is treated as pristine, it remains a societal or natural aberration, rather than an institution resulting precisely from its social-material relations and context. Although markets can be thought of in this way—that is, as a pristine institutional mechanism—doing so entails certain analytical (and perhaps ontological) limitations.

It might be helpful at this point to outline how markets are conceptualized in neoclassical and neoliberal economic thought (see Birch 2017b). According to Dardot and Laval (2014), neoclassical economics tends to frame markets and market competition as a 'condition' in which individual rational decision-making happens; for example, someone like Greg Mankiw (2014: 66)—a Harvard economics professor—defines a market as "a group of buyers and sellers of a particular good or service" who determine the supply and demand of said good or service. Despite being a popular definition, it's obviously pretty basic, theoretically speaking. Dardot and Laval then argue that neoliberals like Hayek frame markets differently, as a 'subjective process' of price discovery—basically markets produce information in the form of prices. These two definitions are very different, one stressing the inputs into economic decision-making (e.g. conditions) and the other stressing the outputs (e.g. prices). A third perspective is also evident in the work of people like Ronald Coase and others theorizing the 'transaction costs' of economic interaction; that is, theorizing markets as the contractual transacting that sits somewhere between the inputs and outputs (Birch 2017b). All these definitions share a common feature though; namely, markets are conceived as abstract mechanisms, free from any sort of social (or natural) arrangement. And, I would argue, this understanding is largely imported into the critical literature on neoliberalism and neoliberal natures (see Chap. 2).

The primary reason for discussing different understandings of markets is to illustrate the extent to which there is debate about what constitutes a market—is it the inputs, the outputs, the transactions, or all of these things? A more relevant issue for my argument in this chapter, however, is the extent to which these abstract notions of markets largely ignore the materialities of those markets. A growing literature on these materialities has emerged at the juncture of economic sociology and anthropology and science and technology studies (STS)—see, for example, the book *Living in a Material World* edited by Pinch and Swedberg (2008). Here, there is increasing interest in understanding how markets are constituted by social actors (e.g. sellers, buyers), social relations (e.g. transactions, valuations), *and* material objects (e.g. cities, food, technologies). As discussed in Chap. 2, key examples of this focus on (political-economic) materiality include the work of people like Donald MacKenzie (2009) on financial markets and Timothy Mitchell (2011) on energy regimes. However, many others, including a number of 'environmental' economic geographers (and others), have also sought to unpack the entanglement of socio-material systems and processes (e.g. Bakker and Bridge 2006; Bridge 2009, 2010; Birch and Calvert 2015; Becker et al. 2016; Birch 2016a; Muniesa et al. 2017). From this perspective, markets are not pristine and untouched abstractions; rather, markets need to be understood and analysed as material relations, institutions, or processes.

Co-Constructing Natures and Markets

Incorporating these insights into my research is a little daunting, as it looks somewhat tricky to integrate these sets of ideas. To start with, it means acknowledging that any analysis of natures and markets necessarily involves combining social and material processes and systems analytically—that is, understanding the world as socio-material, rather than specifically social or material. I've already sought to do this in other research projects on sustainable infrastructure (see Birch 2016a), and my aim is to extend that initial work here. More specifically, in this chapter I want to build on the work of and insights from environmental economic geography and others on the co-construction of environmental and economic processes, systems, and so on (e.g. Hayter 2008; Heidkamp 2008; Hudson 2008; Soyez and Schulz 2008; Patchell and Hayter 2013).

As noted in Chap. 2, this means theorizing both markets and natures as instituted or constructed, although specifically in the interaction with one another rather than in a distinct or abstract form. Nature, as we understand, produce, and live within it, is neither pristine nor abstract, just as markets as we understand, create, and manage them are not pristine nor abstract either. A range of environmental *and* political-economic knowledges, practices, and policies are implicated in this co-construction, including, for example, the development of climate change science, global temperature targets, renewable energy targets, sustainability criteria, labelling and certification of such criteria, and the evaluation of environmental and economic criteria using value chains and/or life-cycle analyses. These 'political-economic materialities', as I am calling them in this book, represent the limits, enablers, and legitimators of specific human action, such as the development of advanced biofuels and the promotion of a bio-economy. In building on Mitchell's (2011) work, my intention is to stress the political-economic, cf. 'political', dimension to and of biophysical materialities—and vice versa—as a means to think about nature-economy relations.

Co-construction can be a tricky conceptual angle to apply, which is important to recognize at the outset of my analysis. Although my aim in the rest of the chapter is to apply this notion of co-construction, I fall back into dichotomies at some points for want of a better theoretical and definitional vocabulary. That being said, I'm going to present the empirical material in a way that illustrates how understanding the development of advanced biofuels (and the bio-economy) from a 'market' *or* 'nature' perspective leaves a gap in understanding the sector, one which the co-constructive angle is meant to address. However, and in order to present this co-constructive analysis, my presentation of the empirical material necessarily adopts a simplified dichotomous account of market and biophysical factors at play in the biofuels sector, and how their necessary co-construction engenders a set of different forms, processes, and outcomes than either of those two factors alone.

THE CO-CONSTRUCTION OF MARKETS AND NATURES IN THE DEVELOPMENT OF ADVANCED BIOFUELS

Analytical and Methodological Framework

My methodological starting point for the analysis in this chapter is the idea that advanced biofuels can be usefully characterized as a (global) value chain (e.g. Ponte 2004; Gereffi et al. 2005; Gibbon et al. 2008; Altenburg 2011; Ponte 2014), representing the various stages in a process of production, consumption, and waste—all part of the analytical underpinnings of environmental economic geography according to people like Gibbs (2006) and Hayter (2008). As these geographers—as well as others—argue, much of mainstream economic geography and political economy tends towards a 'productionist' bias, centring analyses on the production (and labour) process, but ignoring 'feedback' from consumption, use, waste, and disposal. There are generic and specific reasons for going beyond this production-centred approach. On the one hand, nature-economy relations are necessarily about the impacts and implications of both extraction and waste, both being key parts of environmental policymaking (e.g. carbon emissions, carbon sinks) (Jackson 2009). On the other hand, the development of advanced biofuels cannot be understood in terms of production alone, whether we want to analyse their environmental impacts or their political-economic potential; for example, and as I show shortly, the conception of 'waste' and 'residue' is a critical component of any understanding of advanced biofuels.

My research materials for this chapter include in-depth interviews with 20 informants from advanced biofuel value chains in Canada, including private firms, public agencies, and energy experts. Focusing on advanced biofuel value chains—as well as, necessarily, conventional ones—provides an insight into the complex political-economic and material organization of production, consumption, and waste (see Yue et al. 2014); for example, numerous articles, books, reports, and so on about biofuels—and the bio-economy—contain images of one sort or another illustrating bio-based energy and commodity chains (e.g. SDTC 2006; Kedron and Bagchi-Sen 2011; Calvert et al. 2017a; IEA 2017). One example in this book is Fig. 3.3 on the various biofuels pathways. As these images illustrate, there are a series of overlapping issues, concerns, and topics in the political-economic materialities of biofuels. First, the starting point for such analyses is usually an outline of the possible 'technological' pathways for producing biofuels, covering the key conversion technologies (e.g. gasification, hydrolysis, fermentation, transesterification). Second, the technological pathways are often positioned within a broader value or supply chain, including feedstock supply (e.g. grains, lignin, lipids, algae, sugars) and sometimes waste outputs and uses (e.g. distillers grain, heat recovery). Finally, technological pathways and value chains are then sometimes integrated into broader political-economic systems, including sectoral facilities (e.g. refineries, research laboratories), energy landscapes (e.g. pipeline infrastructures, by-product facilities), and institutional support structures (e.g. universities, policy domains, sustainability certifiers and auditors). I've tried to capture each aspect of the overlapping value chains in Fig. 7.1 below, although, no doubt, this is an over-simplification of a much more complex set of interactions and feedback.

The representation of biofuel value chains in this way provides me with a methodological framework to help in the following analysis, primarily because it provides me with a heuristic device for outlining the various stages in biofuel value chains, which I go through below. As such, it should be considered as a methodological representation rather than a 'real-world' description of the value chain. It's worth making, in this regard, some general comments about the specificity of advanced biofuel value chains and their relationship to the bio-economy, at least in the Canadian context. First, when people talk or write about the bio-economy, they frequently focus on the 'downstream' parts of the value chain, by which I mean the stuff that happens after biomass is converted into biofuels or other bio-based products. Moreover, and as I sought to show in Chap. 5, when they

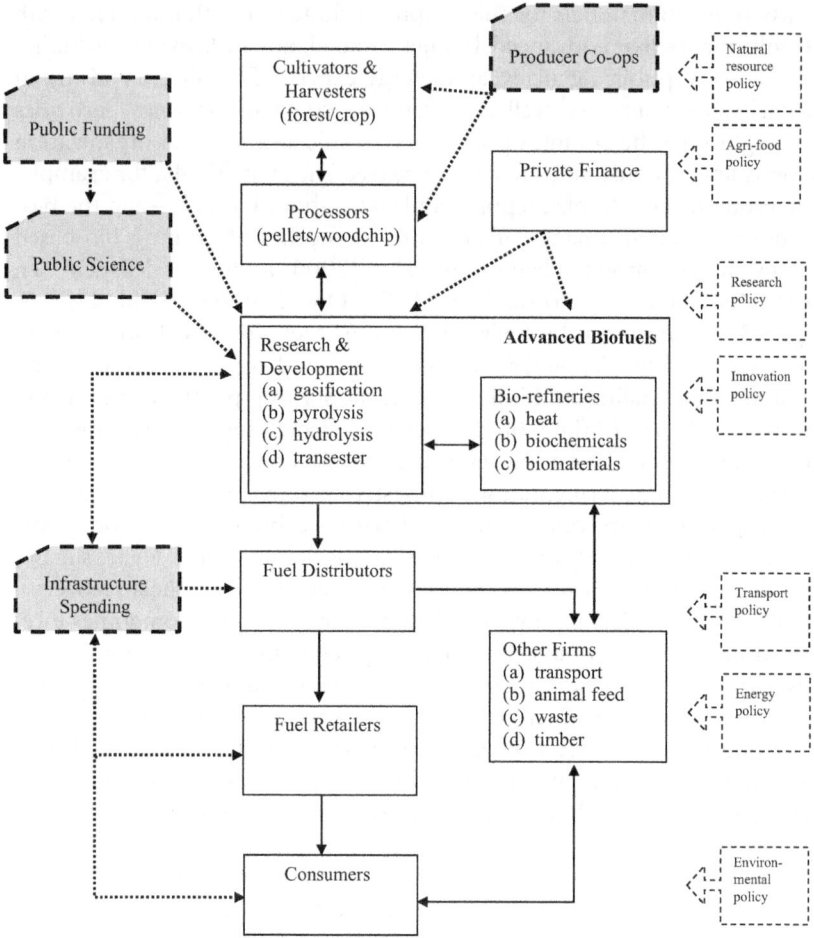

Fig. 7.1 Advanced biofuels value chain

do so, people tend to expect that the downstream value chain will largely stay the same, in that it will just involve a substitution of bio-based inputs for fossil-based inputs. Second, and following on from the last point, people tend to then forget about the 'upstream' value chain, or where all the biomass inputs come from and what's required to get them into shape for conversion into bio-based materials and energy. As a result, biofuels and the bio-economy are often characterized as an easier low-carbon transition than they're likely to be, especially when considering the implications of the political-economic materialities of bio-based outputs (e.g. Birch and Calvert 2015). All of this will become more apparent as I go through the different stages of the value chain below, but it's worth bearing in mind that these political-economic materialities entail outcomes that are often socially, politically, economically, and environmentally problematic, unforeseen, and unexpected.

Upstream Co-construction of Markets and Natures

The starting point for any conventional or advanced biofuel value chain is biomass, and the starting point for biomass is cultivation and harvesting. In the Ontario context, in which I undertook my research, the biomass used to make conventional biofuels was primarily wheat—a high-starch crop—and for advanced biofuels it was primarily wood from forests. I largely focus on the latter in this chapter.

According to several informants, there is (and has been) a considerable quantity of forest biomass available for advanced biofuels, ranging between 40 and 60 percent of the total provincial 'allowable cut'. This 'allowable cut' represents a government-regulated harvestable amount of timber, determined by expectations derived from the knowledges and practices of sustainable forest management, and is not simply a proportion of total forest cover in the province. One informant noted:

> Oh, like in Ontario we're like 40 percent of the annual allowable cut [AAC] and it's not even – you know, we're not even close to the limits. And remember, that like – that singulates over time, right. And the AAC is determined based upon the maximum on an annual basis, right, you know, there's a lot of feedstock out there, but it's just, people don't like the price. (Consultancy #119)

As this comment illustrates, biomass availability, when framed purely in terms of cultivation and harvesting, isn't necessarily a problem in Ontario—there's a significant amount of available biomass—nor does it seemingly contribute to wider problems with food scarcity and food versus fuel debates (Mol 2007; Frow et al. 2009). The importance of considering the political-economic materialities of biomass, then, isn't an issue with its absolute or government-regulated quantity, or even with fears about food scarcity. Rather, biomass availability has to incorporate transport costs, which can sometimes add quite significant financial *and* energy costs to advanced biofuels production, as a direct result of the particular biophysical qualities of timber such as bulkiness and low energy density (Birch and Calvert 2015). As another informant claimed:

> ...the harvesting and movement of these materials any distance obviously requires a lot of energy but also is expensive in terms of manpower and equipment because they tend to – these materials tend to be quite bulky, so not a lot of mass per volume, and not a lot of energy per volume and mass and so moving them any large distance doesn't really make sense which is the why that first bioenergy industry that really off was the pulp and paper industry that could use their own waste as a fuel. So it didn't really have to move the fuel any distance at all because they already had the feedstock's coming in and they already a ready-market for something else. (Energy Expert #130)

As this informant noted, initial bioenergy production largely happened within existing value chains—for example, the pulp and paper industry used the waste from their production processes to generate energy. However, this was only possible because the waste did not need to be transported anywhere, thereby eliminating transport costs from the equation.

As this initial snapshot should show, the political-economic materialities of biomass have ended up configuring specific sectoral value chains in certain ways—that is, pulp and paper firms have used their waste to generate energy for their own use. In the case of advanced biofuels, these sorts of political-economic materialities play out in different ways in reference to market and biophysical dynamics. Next, I address biomass availability first and then turn to feedstock supply second.

Biomass Availability

First, biomass availability—comprising cultivation, harvest, and transport—is dependent on the political-economic configuration and organization of timber resources in Ontario, which differs quite significantly from other countries. Almost 90 percent of land in Ontario is crown or (provincial) government land, as with much of Canada more generally, as this informant outlined:

> The land base for forestry in Canada is about 90% owned by the government in some form or another, either provincial or crown. Provincial crown lands or federal lands that are managed sort of for the public interest by the

government agencies in charge and so that's different than most other countries around the world, including the European ones were split 50/50 or something like that between public and private interests. (Timber Company Manager #115)

As such, the provincial government, which has jurisdiction over natural resources in Canada, currently plays—and will continue to play—a significant role in configuring advanced biofuel value chains. Moreover, conversion technologies—of whatever type—don't remove the geographical constraints imposed by the political-economic materialities of biofuels. Biofuels producers still need access to biomass, and this biomass will be geographically determined (Kedron 2015). In terms of the development of advanced biofuels in Ontario, this is the result of two key political-economic materialities. First, the provincial government enables access to biomass on public lands through timber licensing *and*, more importantly, it effectively subsidizes access to biomass through physical infrastructure support (see Birch 2016a); for example, the government invests in roads so that license holders can access the timber resources, which is something that private land owners simply cannot afford. Second, the collapse of existing value chains, such as pulp and paper, has significant cost implications because it impacts the ability to harvest timber in mixed-species forests, especially relevant since advanced biofuels are easier to produce from softwoods, which I come back to below.

Second, and as the above already implies, biomass availability is also—and perhaps obviously—determined by the biophysical characteristics and geographies of forests and timber. Harvesting timber in Ontario is complicated by the species-mix in Ontario forests, which comprises six to eight commercial varieties—for example, maple, birch, pine, oak, and ash. On top of that, Ontario has four different forest ecosystems spread across the province, although the primary input into advanced biofuel value chains is in Southern Ontario. As several informants noted, different species are used for different things, from flooring through furniture to pulp (and biofuels), and they have different qualities, such as moisture content, knottiness, and so on (Timber Company Manager #115; Advanced Biofuels Producer #124). Here is the kicker: the collapse of existing value chains—see above—that take hardwoods (e.g. pulp and paper industry) means that the cost of harvesting softwoods (e.g. for advanced biofuels production) increases costs, as this informant explained:

And that has two kind of effects, one of them is it increases your per acre cost because, or decreases your per acre revenue, because you're just not cutting as many trees per acre so your volume per acre drops, which means that the road cost and the hours the machines are running to cover the acres stay kind of the same. But the amount of wood that you get out of them is reduced. And so, that, you know, that's a pretty big economic negative. (Timber Company Manager #115)

The biophysical materialities of these different species are further complicated by the different uses and gradings of timber, which I come back to below. These materialities necessitate finding synergies across different sectoral value chains in order to find uses for all the timber harvest, rather than simply substituting an emergent sector (e.g. advanced biofuels) for a declining one (e.g. pulp and paper). Without these synergies—for example, furniture sector taking clear boards, pulp and paper sector taking knotty boards, and biofuels sector taking sawdust and offcuts—then the physical characteristics of forest and timber raise the cost, as two informants pointed out:

Same thing with wood feedstock, high moisture, very high transportation costs in very geographical isolated places, and it makes getting sufficient volumes something that's tricky because it's unusual to be able to rely on a single supplier. And to be able to rely on the entire projects fuel chain or feedstock supply from one source. (Consultancy #135)

So when you go further than that in harvesting trees, you know you end up with the limbs of the trees, you end up with trees that, species that aren't – that can't go into either one of those [potential] products, or you have areas in geography where there really isn't infrastructure in place to make use of that material. (Consultancy #126)

Finally, it is helpful here to examine how these political-economic and biophysical dynamics have played out in Ontario, as a way to analyse the co-construction of markets and natures. The main point to make is that the development of advanced biofuels in Ontario has been (and will continue to be) viable only where it is underpinned by the physical infrastructure and institutions put in place or already established by the provincial government to cultivate, harvest, and transport biomass. The provincial government licenses certain operators to cut timber through long-term licenses that pass down over commercial 'generations', and then supports those operators through the provision of infrastructure to access licensed forests. As such, these operators—in contrast to private woodlot owners—are able to support the continued cutting of mixed-species forests. The reason why is important in light of the arguments of this book and elsewhere about 'neoliberal natures'—namely, the political-economic materialities here limit the impact of market competition (and pricing) on biomass availability, at least according to one informant:

> There's kind of, in a sense, a regulated corporation I suppose and they operate according to guidelines and mandatory regulations that are formulated by the government. And they're kind of independent or more independent of market pricing then private land wood is. (Timber Company Manager #115)

As I've argued elsewhere (e.g. Birch 2016b, 2017b), the conceptualization of neoliberalism as the extension or installation of markets everywhere misses an important aspect of contemporary capitalism; namely, the increasing importance of contract and contract law. Long-term contractual arrangements—as evident here in timber licensing—can protect economic actors from market pressures. Consequently, it is possible to argue that it is the lack of competitive market pressures that provides the necessary condition for the development of innovation pathways like advanced biofuels (Mazzucato 2013).

Feedstock Supply Chains

The second part of the upstream value chains that needs unpacking, especially in light of my claims immediately above, is the feedstock supply chain of advanced biofuels producers. This means going beyond a discussion of forests, timber, and biomass, to consider the specificities of different feedstock supplies and their political-economic materialities.

First, the above discussion of biomass should have demonstrated that there is an increasing supply of biomass for advanced biofuels producers as the previous uses and users of timber decline across Canada. So, as the use of biomass in the pulp and paper sector falls, it frees it up for use elsewhere in the economy—for example, in biofuels production. However, advanced biofuels developers and producers have not pursued this surplus timber for a very simple reason; the political-economic costs of timber harvesting depend on pretty static economies of scale in which there is a primary user (e.g. sawmills) who pays a premium for timber—somewhere between C$125 and C$150 per 'bone dry' metric ton—and then secondary and tertiary users who pay different price points depending on the quality of leftover timber. As this informant explained:

So if you're primary – if you can't find a feedstock that has, like as a second-ary basis, that actually has a lower potential economic value [than primary use], even though the tree is always the same price to cut it and harvest it, it's a question of who's extracted the best value out of those trees. So in the older days, if you think of it, veneer was the primary product that came out of poplar trees, right? So they would pay a premium. So the next guy on the use of the tree would maybe be a pallet sawmill and then the next guy in the tree would be you know OSB [oriented strand board], or pulp, right. But the price points were different, even though the cubic metre in the tree itself, you know the quality and form as its merchandised, has to bear differ-ent price components in order to make it affordable. (Consultancy #126)

As a result, advanced biofuels producers are not viable primary users of timber because the cost structure simply militates against the production of an economically viable biofuel (e.g. one that people will buy instead of petroleum/gasoline). Consequently, advanced biofuels developers and producers in Canada—and, no doubt, elsewhere in the world—are focused on identifying a feedstock that has little or no 'value' in existing markets. Unlike ethanol where feedstock is the major cost, the most significant cost in advanced biofuels production is the technology (e.g. enzymes), at least according to several informants (e.g. Consultancy #119). This means advanced biofuels developers and producers are largely trying to find ways to convert waste, leftovers, offcuts, and sawdust into a liquid biofuel, rather than planks. As this informant commented:

However, they [sawmills] remain very concerned that these pulp and paper facilities that they're beholden to for their costs – they cover their costs basically. They need to find someplace else for that residual product to go. So these advanced biofuels are really changing – have the potential to really change the game on the whole energy front – the whole interdependency of the forestry market. (Non-profit #116)

Second, the biophysical materialities at play centre on the search for an 'agnostic' feedstock as the biological material input base for producing advanced biofuels—as well as other bio-based products in a bio-economy. Several informants made comments to this effect (e.g. #115, #116, #135). It makes sense, in light of the discussion above about biomass availability, to focus on finding a technological process that can convert any feedstock into a liquid fuel, since it would solve a range of problematic issues—I come back to this below. Advanced biofuels are currently premised on the idea that there are plentiful quantities of 'wasted' feedstock out there available to producers, although this has turned out to be largely illusory, as this informant noted:

> And I think there was deeper realisations that the cost and logistics associated with feedstocks, which is much more complicated and difficult and ultimately higher cost than what people thought. And so the higher cost of feedstock plus the lack of cost reductions in the conversation process, you know, kind of stalled out the development of cellulosic ethanol. (Fossil Energy Producer #118)

A key issue here, however, is how to address the biophysical challenges of sourcing, transporting, and processing heterogenous feedstock. Just a few examples of these challenges include the long rotation periods for forests compared with other feedstock, the fact that sawdust is explosive and difficult to transport as a result, and the issue of taking forest residues since they are important for nutrient recycling.

Finally, the co-construction of markets and natures in feedstock supply becomes evident in the framing of offcuts, leftovers, sawdust, and suchlike as 'residues', an economic *and* material category that avoids the assumption that advanced biofuels producers are 'primary' timber users—and therefore primary payer for forest harvesting—while reinforcing the idea that advanced biofuels are derived from natural resources that have no value (e.g. waste, offcuts, etc.) and are, coincidentally, sustainable and renewable. However, this 'residues' category is dependent on plugging advanced biofuels into existing value chains (e.g. sawmills), rather than replacing or substituting for them. Moreover, advanced biofuels—as well as other bio-based sectors—seem to be dependent on this framing of residues as valueless and free to use—a "costless raw material", according to one informant (Conventional Biofuels Producer #125). From a political ecology perspective, we could criticize yet another assumption that natural resources are simply 'out there' and free to use (e.g. Foster 2002). However, a more interesting point to note is that residues—and the economic viability of advanced biofuels—are based on the use of a feedstock precisely because it's *not* a competitive market:

> With relatively few markets [for sawdust], it's not a really competitive marketplace. And the other thing is, what makes it particularly attractive in terms of bioenergy is that you're converting a non-edible bioproduct as opposed to, like, corn ethanol. The argument is always made that we shouldn't be using food to drive our cars. (Advanced Biofuels Producer #124)

Several other contradictions arise in light of this analysis with the 'neoliberal natures' understanding of nature-economy relations (see Castree 2008). First, advanced biofuels production depends on mitigating feedstock risk and securing constant feedstock supply (Conventional Biofuels Producer #125; Consultancy #135), suggesting, again, that long-term contractual arrangements—rather than market(-like) ones—are central here. Second, the development and expansion of advanced biofuels—and other bio-based sectors—will necessarily lead to the commodification of so-called 'residues' as these residues are made valuable; for example, one informant stated that demand for residues "gives them [sawmills] an alternative or an additional market" (Advanced Biofuels Producer #102). As such, advanced biofuels potentially contain the seeds of their own eventual failure within their own political-economic materialities, since residues will gain value as demand for them rises.

Downstream Co-construction of Markets and Natures

Rather than try to cover every stage of the biofuels value chain—such as 'pre-treatment' of biomass, meaning the densification of bulky biomass into less dense commodity with higher energy return per ton—I'm going to focus on certain points in the value chain. In this sub-section, I start with technology conversion processes and then turn to distribution and retail.

Technology Conversion

The pivot, as it were, in advanced biofuel value chains is the conversion technology researched, developed, and deployed by advanced biofuels producers—as well as other firms developing other bio-based products (e.g. biochemicals, bioplastics, etc.). The above analysis covers the upstream parts of the value chain, while the below analysis covers the downstream (see Fig. 7.1). The first thing to note in relation to advanced biofuels developers and producers is the range of conversion technologies used to transform biomass, of many different kinds, into liquid fuels. As James Smith (2010) notes, the idea of liquid biofuel has been around for at least 100 years, while gasification conversion technologies like Fischer-Tropsch synthesis were developed in the 1920s (WorldWatch Institute 2007). Across the different conversion technologies, however, there's a general concern with finding ways to convert different and varied feedstock into a liquid fuel form, whether through different forms of hydrolysis, pyrolysis, or gasification (ibid.: 63–66). As a result, advanced biofuel value chains are different from conventional biofuels, requiring new kinds of logistics matching the new technologies (e.g. Ministry Agency #133).

As mentioned above, advanced biofuels producers are focused on developing conversion technologies that are 'feedstock agnostic' so that they can integrate any potential feedstock—especially cheap 'residues'—into their production. An informant even suggested that "the essence of the value chain, if you're looking at the fuel, it only starts at the residuals" (Advanced Biofuels Producer #102). While this ignores potential changes in the upstream value chain (e.g. changes to the collection and trucking of residues), it does illustrate the emphasis on finding a technology pathway that is capable of converting varied feedstock. This is very different from conventional biofuels, as one informant noted:

> So some of the first generation biofuel had a very narrow range of their feedstock. And I mean you probably heard this already but their processes – the advanced biofuels now they want to become feedstock agnostic. So that's the real key to changing how they process their – what they can accept for feedstock and all that kind of stuff really does change the game on especially the geography I would say. (Non-profit #116)

The reason for this is quite simple; the potential feedstock supply increases significantly if a conversion technology is 'agnostic', opening up more waste, offcuts, and leftovers to be transformed into a 'residual' resource for eventual conversion into biofuels. Similar concerns are evident in the discussions about using so-called 'marginal land' as new cultivation sites, especially in the Global South (e.g. Bailis and Baka 2011; Baka 2014). One informant framed the development of conversion technologies as "looking at a means to be able to deliver – or to in effect valorize waste" (Advanced Biofuels Producer #114), although it's important to emphasize that this means valorizing it by producing biofuels, not by selling waste.

Secondly, the biophysical materialities of feedstock drive the development of conversion technologies towards feedstock agnosticism, where possible. As already mentioned above, Ontario's forest biomass is diverse, encompassing six to eight main tree species including both softwoods and hardwoods. Furthermore, there are other biophysical factors at play, including moisture content and chemical quality, which all impact the use of trees. First, softwoods and hardwoods have different moisture contents, which "will run anywhere between 55 and 63 percent [for softwoods]. Hardwood will get anywhere from 42 to 55" according to one informant

(Advanced Biofuels Producer #124). Second, moisture content is not the only factor determining feedstock preferences, as the same informant noted when it comes to preferring softwoods over hardwoods:

> It's the difference – different – you end up with a different quality product and our product has to meet a whole bunch of very stringent standards in terms of emissions, and we're more easily able to obtain those – reach those standards with red pine white pine than we are with hardwood. Softwoods are preferred to hardwoods. (Advanced Biofuels Producer #124)

The chemical qualities of wood can be affected by a range of things, including the age of the tree, the part of the tree, and so on. All of this is complicated further by market designations like the four to five timber gradings used to define timber as 'board', 'knotty', 'offcut', 'sawdust', and so on (Timber Company Manager #115). Taking all these things into account—that is, species, moisture, chemistry, and gradings—means that there are at least 20-odd timber products or residues coming out of forests alone. Understanding these biophysical dynamics helps us understand the drive behind feedstock agnosticism of advanced biofuels producers, at least in the Canadian context—other contexts will likely be configured by different natural resource endowments (Harvey and Pilgrim 2011).

Finally, the co-construction of markets and natures in conversion technologies is not only reflected in the simultaneous aim of developing feedstock agnostic conversion technologies—that can account for the different market and biophysical characteristics of diverse feedstock—it is also evident in the aim of producing a homogenous commodity (i.e. sugars) that can be used by a range of potential customers. One informant put it like this:

> …you [were] mentioning that other people that you spoken to expressed the desire to turn all kinds of cellulosic materials into sugars so that they were comparable products. That is the goal, commodifying or turning wood into a commodity or turning cellulosic, everything into a commodity so that it can't be differentiated, so it can be marketized and be moved around. (Consultancy #135)

Such homogenous commodities are defined as 'fungible' in political-economic literature, meaning that the commodity (e.g. sugar) is perfectly interchangeable no matter who produces it. Here, the co-construction of markets and natures entails the development of conversion technologies that can treat diverse inputs as effectively fungible—so that it doesn't matter what feedstock you use—and which, in turn, produce a fungible commodity—so that it doesn't matter who produces the resulting output (e.g. sugars). While this could be seen as a (neoliberal) form of commodification of nature (Castree 2006, 2008), there is something more complex going on here in my view. On the one hand, a diverse range of biological matter—itself differentiated through political-economic knowledge claims (e.g. timber gradings)—is treated as an 'agnostic' input through conversion technologies. On the other hand, a homogenous output (i.e. sugars) is produced by these conversion technologies. As one informant argued (Advanced Biofuels Producer #114), "It comes down to cost and fungibility of the product" in that:

Respondent: Fungibility basically means, right, that I know if I buy a gallon of ethanol from Bob and I buy a gallon of ethanol from Bill, I know it's going to be the same ethanol.
Interviewer: Yeah.
Respondent: And I know in a very transparent form what that market looks like from Bill or Bob.

Here, the 'market' is equated with fungibility—or political-economic *and* biophysical homogeneity—in that all inputs and all outputs of advanced biofuels producers are framed and treated as the same. However, and this is a key problem with advanced biofuels at present, these advanced biofuels are not fungible within current infrastructure. According to the same informant, "the infrastructure of getting ethanol to market is pretty straightforward", which means that "it's very, very fungible", while, and in contrast, the "fungibility of the feedstocks of waste say versus other second-generation feedstocks are less defined" (Advanced Biofuels Producer #114). It's not enough, then, to develop feedstock agnostic conversion technologies, since fungibility depends on a broader array and configuration of social (e.g. commodity trading data) and material (e.g. tankers) infrastructures that stretch both upstream and downstream, for example, paying farmers and foresters to collect residues (Consultancy #119).

Distribution and Retail
A key element in the advanced biofuels value chain is distribution and retail. This isn't only for the obvious reason that any product needs to be sold; rather, it's also because of continuing issues with developing and producing a viable advanced biofuel—by which I mean an economical one (see Chap. 3). Almost any biological matter can be converted into a liquid

fuel with the right technology, but whether that conversion process is economical is the crux of the matter. As before, determining the 'economicality'—for want of a better word—of a biofuel entails an examination of market and biophysical dynamics.

First, the distribution and retailing of advanced biofuels is still tied to specific jurisdictional mandates. For example, Ontario's 5 percent renewable fuels mandate, introduced with the 2007 Ethanol in Gasoline regulation, treats 1 litre of cellulosic biofuels as equivalent to 2.5 litres of conventional ethanol. In 2017, the provincial government decided to extend the mandate to 10 percent in 2020. The biofuels market in Ontario, as well as Canada, is seemingly built on the back of this political-economic decision, since market prices aren't enough to support the development and expansion of advanced biofuels. Current attempts, then, to promote renewable liquid fuels do not necessarily involve 'market-like' rules— which Peck and Tickell (2002) argue are characteristic of neoliberalism— although they might reflect a broader, more Polanyian notion about markets as 'instituted process' (Polanyi 1957), although only if, and this is crucial, markets actually emerge and prosper. At present, however, advanced biofuels appear to bolster incumbent industrial sectors—see Chap. 5—rather than compete or disrupt them. As one informant argued:

> And the way they've [biofuels] been deployed is primarily with ethanol as a fuel additive to gasoline and with bio-diesel as a fuel additive to diesel. So, in effect, by having these biofuels just being an additive to the existing fuel system, they've in fact reinforced the current infrastructure that we have. (Energy Expert #27)

In light of the discussion above about creating fungible commodities— and their fungibility being reliant on the underlying socio-material infrastructures—it doesn't look like advanced biofuels will end up as a standalone market.

Second, in terms of the biophysical dynamics at the distribution and retail points in the advanced biofuel value chain, there are a range of issues that configure the development of advanced biofuels. A critical consideration is that distribution—just like feedstock supply—is determined by the certain geographical materialities (see Birch and Calvert 2015; Calvert et al. 2017a, b). In the Canadian context, for instance, the USA looks like a huge potential market; however, several informants stressed that the location of feedstock determine where the market *could* be, in that advanced biofuels producers have to acquire their feedstock from a 100-mile radius from their production facility *and* then sell into an energy market that is limited to 100-mile radius (Conventional Biofuels Producer #125). Co- or by-products face similar limitations, since biofuels producers—usually smaller firms—do not have the capacity to access larger, more geographically distributed markets. Two informants made similar points in this regard:

> ... they're [companies] realising that transportation fuels are really not highly valued. It's a very large market, but when you're a start-up company, who gives a crap how big the market is? You're just trying to make it through the next quarter. (Consultancy #119)

> And I think there was deeper realisations that the cost and logistics associated with [advanced biofuel] feedstocks, which is much more complicated and difficult and ultimately higher cost than what people thought. And so the higher cost of feedstock plus the lack of cost reductions in the conversation process, you know, kind of stalled out the development of cellulosic ethanol. (Fossil Energy Producer #118)

These biophysical limits result from the material qualities of feedstock and biofuels, including: first, the bulkiness of biomass and the need to truck it from harvest site to refinery, rather than send it (easily) by ship, rail, or pipeline, and second, the chemical qualities and lower energy density of biofuels compared with fossil fuels, resulting from water content and potential evaporation if piped and the resulting need for trucking it again (Calvert et al. 2017a).

Finally, the co-construction of markets and natures here involves the synergies between biofuels and other commodity markets, including timber, animal feed, and other co-products. Co-construction entails the coordination of biofuel value chains, infrastructures, and customer base with the specific feedstock characteristics. More specifically, different feedstocks entail different 'landscapes' (Consultancy #135) which have particular environmental economic geographies. For example, the configuration of these landscapes differs between conventional and advanced biofuels, as well as between different types of advanced biofuels. First, corn is very much a 'rural' feedstock configured by the interaction between biophysical materialities and political-economic infrastructures; for example, ethanol cannot go into pipelines due to its oxygen content, as this informant explained:

> So from an infrastructure standpoint both ethanol and biodiesel are blended at a blending terminal and then trucked to retail. ... So ethanol and biodiesel because they have some oxygen in them and some clean

properties associated with their structure and the fact that they've got some oxygenation in their compounds, we tend to clean pipelines out so they don't tend to be shipped by pipeline as neat products or as blends. Especially in North America where you've got long, long pipelines. It's done somewhat in Europe but our pipelines here in Canada and the US tend to be very long, so from an infrastructure standpoint. (Consultancy #123)

For this reason, ethanol in Canada is primarily sold within a localized market—within 100 miles—in order to cut down on transport costs (Conventional Biofuels Producer #125). However, as this quote also indicates, this is a consequence of the specific infrastructure geographies in Canada, and North America, where there are longer energy pipelines. Second, advanced biofuels landscapes are configured differently, depending on their specific political-economic materialities. For example, cellulosic biofuels are derived from 'residues' (see above), necessitating the wholesale installation of new value chains and infrastructures to collect, process, and distribute feedstock (Consultancy #119); in contrast, waste-based advanced biofuels have an established infrastructure and value chain already in place, as this informant noted:

In terms of key infrastructures, I would say the waste industry – the value chain for waste to biofuels is – is fairly established. You know, people come by every day throughout the world to pick trash up at your curb, right. (Advanced Biofuels Producer #114)

The point of this discussion here is to stress that the particular political-economic materialities of different biofuels are often geographically specific; that is, existing and potential feedstock value chains and infrastructure configure biofuels markets. The implications of this are that some places have little chance to develop a bio-based economy because they simply lack the feedstock base (e.g. no corn, no forests, no coast) or the infrastructure necessary to integrate feedstock into energy production (e.g. roads, trucks, processing plants).

Cross-Cutting Co-construction of Markets and Natures

It is worth ending this discussion by considering two cross-cutting aspects of biofuel value chains that are difficult to clearly differentiate into one stage or another; namely, financing and sustainability. I can only briefly outline these two issues here, but it's worth stressing that finance and sustainability are two key driving concerns in the development of advanced biofuels.

First, financing can be split between innovation and investment financing. Innovation financing comes from a range of sources with different underpinning rationales (Birch 2017c), including government schemes (e.g. research priorities) and private investors (e.g. venture capital); in contrast, investment financing is primarily driven by the cost structure of things like feedstock and capital expenditure, even when there is government support (e.g. Canada's NextGen Biofuels Fund). As should be obvious from reading this book, in my view the main reason that advanced biofuels have not taken off as expected—despite all the mandates and policy support—is because of problems attracting investment financing, especially as the result of the uncertainty around the cost structure of biofuels refineries, which can cost several hundred million dollars (Kedron 2015). As noted already, advanced biofuels entail significant technology costs (e.g. R&D, enzymes, energy, etc.), which means they are reliant on low feedstock costs. Advanced biofuels, therefore, have to be integrated into other bio-based value chains as they cannot afford to be primary feedstock users. Analysing this dynamic in terms of the co-construction of markets and natures means understanding financing as constituted by the combining of biofuels with other bio-based products, whether they are 'low-tech' products (e.g. furniture) or 'high-tech' products (e.g. bioplastics). Working out how to finance this setup, however, is tricky because it requires a form of systemic and synergistic financial investment (e.g. financing the whole value chain), rather than discrete investment (Stephen et al. 2012). Consequently, the political-economic materialities of advanced biofuels may militate against specific forms of private investment, necessitating more direct involvement of the state in supporting their expansion (Mazzucato 2013).

Second, sustainability is critical part of the advanced biofuels puzzle as well. As discussed in Chap. 4, in the Ontario context, there is often a tendency to equate the use of renewable biomass (e.g. corn, trees) with 'sustainable' business practices and markets. As one informant noted (Consultancy #135), however, while residues, trees, and waste are treated as sustainable in Ontario, this can ignore the problematic qualities (e.g. chemical content) of residues like varnished wood, wood from transfer stations, and suchlike. As several other informants stressed to me during my research, sustainability can be—and is—measured in different ways (Energy Expert #27), which is highlighted in the literature as well (e.g. Scarlat and Dallemand 2011). Increasingly, life-cycle analyses (LCA) are used to measure sustainability because they enable the assessment of various inputs and outputs throughout the value chain. While this requires a dedicated chain of custody approach in order to trace the origin, use, and disposal of biomass (Non-profit #116), it also entails an epistemic decision regarding the system boundary of any value chain. As one informant pointed out, for example:

> But then you get into these really tricky, thorny questions about where you draw the system boundaries, what do you include in the production of bio-fuels? So one example of that is do you even include the food required to feed the labor workers who work on the field, or not? And if the answer is yes, then must you do the same thing for petroleum refining? (Energy Expert #131)

Measurement processes like LCA are always relative, in that their components require weighting, which highlights the key issue with the sustainability of advance biofuels. Sustainability criteria (e.g. GHG emissions reductions) cannot be embodied in the biophysical qualities of the fuel itself—unlike performance criteria—which means that it's difficult to get international agreement (or even national or regional agreement) on sustainability standards. As this informant argued:

> But as mentioned, if we look forward to really doing something with climate change and expanding the biofuels to higher levels, at some point we do need to have some kind of policies in place for sustainability and I think everyone recognizes that. So I don't think anyone's opposed to having the discussion at this juncture. It's more filtering through to – to the real meat of the issue and – and trying to find some way to try to separate the political from – from the more technical or practical issues. Or arguments or what-ever. (Consultancy #123)

CONCLUSION

Advanced biofuels have particular political-economic materialities that configure their development, deployment, and uptake. At present, advanced biofuels are struggling to progress beyond research, development, and pilot or demonstration stages, for a number of reasons I've tried to unpack in this chapter (and in the rest of the book). My main aim here has been to analyse the co-construction of markets and natures as a theo-retical framework for understanding advanced biofuels. Using a value chains approach, I examined upstream and downstream stages in the value chain, as well as two cross-cutting issues: finance and sustainability. For the analysis of the upstream, covering biomass availability and feedstock supply, and downstream—covering conversion technology and distribution—stages I first discussed the political-economic and material dynamics before outlining how these dynamics can be understood within the analytical framework based on the co-construction of markets and natures.

My discussion of the co-construction of markets and natures during these stages leads to several implications worth considering. First, the development and deployment of advanced biofuels will lead to significant changes in the upstream value chain; we cannot simply assume that the current state of affairs will continue (see Chap. 5). For example, and unlike fossil fuel resources, biomass will involve distributed and decentralized collection from a number of often small woodlots, or from government-managed woodlots; the need for onsite densification and bulk shipping will necessitate significant reshaping of landscapes (e.g. wider roads, more trucks). Second, advanced biofuels are currently based on finding cheap feedstock supplies, especially through the use of 'waste' and 'residues'. However, 'residues' are not likely to remain freely or cheaply available for biofuels producers, since residues will acquire value as demand for them grows. How this will impact biofuels producers in the long-run is critical and necessitates avoiding market competition altogether (e.g. long-term contracts). Third, developing 'agnostic' conversion technologies and fungible commodities entails more than a simple techno-fix. Rather, fungibility depends on the configuration of socio-material relations, such that something (e.g. feedstock) can be framed and integrated as fungible within socio-material infrastructures. Finally, some form of ideal or abstract market is not installed or inserted into these nature-economy relations, as the neoliberal natures might suggest. 'Markets'—for want of a better word—might emerge through the interaction between economic and environmental processes, but they are frequently bounded by state institutions, framed by naturalistic discourse, and enacted through non-market practices (e.g. contractual arrangements).

REFERENCES

Altenburg, T. (2011) Interest groups, power relations, and the configuration of value chains: The case of biodiesel in India, *Food Policy* 36: 742–748.

Bailis, R. and Baka, J. (2011) Constructing sustainable biofuels: Governance of the emerging biofuel economy, *Annals of the Association of American Geographers* 11(4): 827–838.

Baka, J. (2014) What wastelands? A critique of biofuel policy discourse in South India, *Geoforum* 54: 315–323.

Bakker, K. and Bridge, G. (2006) Material worlds? Resource geographies and the 'matter of nature', *Progress in Human Geography* 30(1): 5–27.

Becker, S., Moss, T. and Naumann, M. (2016) The importance of space: Towards a socio-material and political geography of energy transitions, in L. Gailing and T. Moss (eds) *Conceptualizing Germany's energy transition*, London: Palgrave Macmillan, pp. 93–108.

Birch, K. (2006a) The neoliberal underpinnings of the bioeconomy: The ideological discourses and practices of economic competitiveness, *Genomics, Society and Policy* 2(3): 1–15.

Birch, K. (2006b) Introduction: biofutures/biopresents, *Science as Culture* 15(3): 173–181.

Birch, K. (2016a) Materiality and sustainability transitions: Integrating climate change into transport infrastructure in Ontario, Canada, *Prometheus: Critical Studies in Innovation* 34(3–4): 191–206.

Birch, K. (2016b) Market vs. contract? The implications of contractual theories of corporate governance to the analysis of neoliberalism, *Ephemera: Theory & Politics in Organization* 16(1): 107–133.

Birch, K. (2017a) Introduction: Techno-economic assumptions, *Science as Culture* 26(4): 433–444.

Birch, K. (2017b) *A research agenda for neoliberalism*, Cheltenham: Edward Elgar.

Birch, K. (2017c) Financing technoscience: Finance, assetization and rentiership, in D. Tyfield, R. Lave, S. Randalls and C. Thorpe (eds), *The Routledge handbook of the political economy of science*, London: Routledge, pp. 169–181.

Birch, K. and Calvert, K. (2015) Rethinking 'drop-in' biofuels: On the political materialities of bioenergy, *Science and Technology Studies* 28: 52–72.

Borup, M., Brown, N., Konrad, K. and van Lente, H. (2006) The sociology of expectations in science and technology, *Technology Analysis & Strategic Management* 18(3–4): 285–298.

Bridge, G. (2009) Material worlds: Natural resources, resource geography and the material economy, *Geography Compass* 3(3), 1217–1244.

Bridge, G. (2010) Resource geographies 1: Making carbon economies, old and new, *Progress in Human Geography* 35: 820–834.

Calvert, K., Birch, K. and Mabee, W. (2017a) New perspectives on an old energy resource: Biomass and emerging bio-economies, in S. Bouzarovski, M. Pasqualetti and V. Castan Broto (eds) *The Routledge research companion to energy geographies*, London: Routledge, pp. 47–60.

Calvert, K., Kedron, P., Baka, J. and Birch, K. (2017b) Geographical perspectives on sociotechnical transitions and emerging bio-economies: Introduction to a special issue, *Technology Analysis & Strategic Management* 29(5): 477–485.

Castree, N. (2006) Commodifying what nature? *Progress in Human Geography* 27(3): 273–297.

Castree, N. (2008) Neoliberalising nature: The logics of deregulation and reregulation, *Environment and Planning A* 40: 131–152.

Dardot, P. and Laval, C. (2014) *The new way of the world*, London: Verso.

Foster, J.B. (2002) *Ecology against capitalism*, New York: Monthly Review Press.

Frow, E., Ingram, D., Powell, W., Steer, D., Vogel, J. and Yearley, S. (2009) The politics of plants, *Food Security* 1(1), 17–23.

Gellert, P.K. (2005) For a sociology of 'socionature': Ontology and the commodity-based approach, in P. Ciccantell, D. Smith, and G. Seidman (eds) *Nature, raw materials, and political economy (Research in rural sociology and development, volume 10)*, Bingley: Emerald Group Publishing Limited, pp. 65–91.

Gereffi, G., Humphrey, J. and Sturgeon, T. (2005) The governance of global value chains, *Review of International Political Economy* 12(1): 78–104.

Gibbon, P., Bair, J. and Ponte, S. (2008) Governing global value chains: An introduction, *Economy and Society* 37(3): 315–338.

Gibbs, D. (2006) Prospects for an environmental economic geography: Linking ecological modernization and regulationist approaches, *Economic Geography* 82(2): 193–215.

Harvey, M. and Pilgrim, S. (2011) The new competition for land: Food, energy, and climate change, *Food Policy* 36: 540–551.

Hayter, R. (2008) Environmental economic geography, *Geography Compass* 2: 1–20.

Heidkamp, P. (2008) A theoretical framework for a 'spatially conscious' economic analysis of environmental issues, *Geoforum* 39: 62–75.

Heynen, N., Kaika, M. and Swyngedouw, E. (eds) (2006) *In the nature of cities*, London: Routledge.

Hilgartner, S. (2007) Making the bioeconomy measurable: Politics of an emerging anticipatory machinery, *BioSocieties* 2: 382–386.

Hudson, R. (2008) Cultural political economy meets global production networks: A productive meeting?, *Journal of Economic Geography* 8(3): 421–440.

IEA. (2017) *Technology roadmap: Delivering sustainable bioenergy*, Paris: International Energy Agency.

Jackson, T. (2009) *Prosperity without growth*, London: Earthscan.

Kedron, P. (2015) Environmental governance and shifts in Canadian biofuel production and innovation, *The Professional Geography* 67(3): 385–395.

Kedron, P. and Bagchi-Sen, S. (2011) A study of the emerging renewable energy sector within Iowa, *Annals of the Association of American Geographers* 101(4): 882–896.

MacKenzie, D. (2009) *Material markets: How economic agents are constructed*, Oxford: Oxford University Press.

Mankiw, G. (2014) *Principles of economics*, Independence, KY: Cengage Learning.

Mazzucato, M. (2013) *The entrepreneurial state*, London: Anthem Press.

McKibben, B. (2012) Global warming's terrifying new math, *Rolling Stone* 19 July, retrieved from: https://www.rollingstone.com/politics/news/global-warmings-terrifying-new-math-20120719.

Mitchell, T. (2011) *Carbon democracy*, London: Verso.

Mol, A. (2007) Boundless biofuels? Between environmental sustainability and vulnerability, *Sociologia Ruralis* 47(4): 297–315.

Moore, J.W. (2015) *Capitalism in the web of life*, London: Verso.

Muniesa, F., Doganova, L., Ortiz, H., Pina-Stranger, A., Paterson, F., Bourgoin, A., Ehrenstein, V., Juven, P., Pontille, D., Saraç-Lesavre, B. and Yon G. (2017) *Capitalization: A cultural guide*, Paris: Presses de Mines.

Patchell, J. and Hayter, R. (2013) Environmental and evolutionary economic geography: Time for EEG²?, *Geografiska Annaler B* 95(2): 111–130.

Peck, J. and Tickell, A. (2002) Neoliberalizing space, *Antipode* 34(3): 380–404.

Pinch, T. and Swedberg, R. (eds) (2008) *Living in a material world*, Cambridge, MA: MIT Press.

Polanyi, K. (1957) The economy as instituted process, in K. Polanyi, C. Arensberg and H. Pearson (eds) *Trade and market in the early empires*, Illinois: Free Press and Falcon's Wing Press, pp. 239–270.

Ponte, S. (2004) *Standards and sustainability in the coffee sector*, Manitoba: International Institute for Sustainable Development.

Ponte, S. (2014) The evolutionary dynamics of biofuel value chains: From unipolar and government-driven to multipolar governance, *Environment and Planning A* 46(2): 353–372.

Richardson, B. (2012) From a fossil-fuel to a biobased economy: The politics of industrial biotechnology, *Environment and Planning C* 30: 282–296.

Robbins, P., Hintz, J. and Moore, S. (2010) *Environment and society*, Oxford: Wiley-Blackwell.

Scarlat, N. and Dallemand, J-F. (2011) Recent developments of biofuels/bioenergy sustainability certification: A global overview, *Energy Policy* 39: 1630–1646.

SDTC. (2006) *SD business case: Renewable fuel – Biofuels*, Ottawa: Canada Foundation for Sustainable Development Technology.

Smith, J. (2010) *Biofuels and the globalization of risk*, London: Zed Books.

Smith, N. (1984[2008]) *Uneven development*, Athens, GA: University of Georgia Press.

Soyez, D. and Schulz, C. (2008) Editorial: Facets of an emerging environmental economic geography (EEG), *Geoforum* 39: 17–19.

Stephen, J., Mabee, W. and Saddler, J. (2012) Will second-generation ethanol be able to compete with first-generation ethanol? Opportunities for cost reduction, *Biofuels, Bioproducts & Biorefining* 6: 159–176.

Stern, N. (2010) *A blueprint for a safer planet*, London: Vintage Books.

Tickell, A. and Peck, J. (2003) Making global rules: Globalisation or neoliberalisation?, in J. Peck and H. Yeung (eds) *Remaking the global economy*, London: SAGE, pp. 163–182.

Wood, T. and Mabee, W. (2016) *ACW: Baseline report – Energy*, Working Paper #ACW-04, York University.

WorldWatch Institute. (2007) *Biofuels for transport*, London: Earthscan.

Yue, D., You, F. and Snyder, S. (2014) Biomass-to-bioenergy and biofuel supply chain optimization: Overview, key issues and challenges, *Computers and Chemical Engineering* 66: 36–56.

CHAPTER 8

Conclusions: Alternative Bio-Economies

INTRODUCTION

The cost of renewable energy production is falling every year, across the range of energy sources—including bioenergy (Romm 2018). As a result, there's an increasing expansion of renewable energy production, although it hasn't replaced—nor is it anywhere near replacing—fossil energy production, at least in most countries. The potential of renewable energy to reduce carbon emissions ensures that there's an ongoing support and investment in its expansion, yet fossil fuels remain the dominant energy regime across most of the world. Why? For people like David Tyfield (2014, 2017), energy resources like coal—one of the worst carbon emitters—remain a major energy source precisely because of "the strong connections between coal-based socio-technical systems and a political regime of classical liberalism" (Tyfield 2014: 59). From this perspective, while our current political-economic regime—which is frequently characterized as 'neoliberal'—remains dominant, we are unlikely to witness or be able to pursue a system-transforming socio-technical transition towards renewable energy, or a bio-economy. Not that long ago—shortly after the 2007–2008 global financial crisis—there was some hope that the end of neoliberalism was nigh (e.g. Birch and Mykhnenko 2010) and therefore that such system transformation, towards a 'green new deal' for example, might be possible (Goldstein and Tyfield 2017). However, that now seems like a false dawn (Mirowski 2013), meaning that we can expect fossil energy to remain the dominant energy regime into the foreseeable future.

© The Author(s) 2019
K. Birch, *Neoliberal Bio-Economies?*,
https://doi.org/10.1007/978-3-319-91424-4_8

And yet, if we problematize 'neoliberalism' as a concept and a way to characterize our current political-economic regime, as I've been arguing throughout this book and elsewhere (Birch 2016, 2017), then we might find it easier to support and promote the system transformation that most critical environmental thinkers deem necessary to stave off climate disaster (e.g. Jackson 2009; McKibben 2012, 2016; Berners-Lee and Clark 2013; Moore 2015; Bonneuil and Fressoz 2017). I've attempted to do just that throughout this book, probably with differing degrees of success. Whether or not you agree with my argument, it's still important to consider the different ways that academics, politicians, policy-makers, activists, and others can and do understand nature-economy relations; this is especially the case if we think that ideas, visions, and discourse are, in whatever way, somehow performative or constitutive of socio-natural reality. For example, if natural resources are not 'natural' as Bridge (2009) argues, then we have to understand how something like biomass is turned into a resource through political-economic processes—which could include the extension of markets—as well as transformations of socio-technical regimes and (potentially utopian) visions of future societies we want. Crucially, I've been trying to stress throughout most of the preceding chapters that this sort of analysis also needs to take biophysical materiality into account in a way that means it's not simply counterposed to one political-economic process or another (e.g. privatization). Rather, there's a need for a conceptual approach underpinned by the idea of 'political-economic materiality' and what this means for the bio-economy.

With all this in mind, my main aim in this concluding chapter is to discuss alternative conceptions of the bio-economy, ones which are often sidelined or hidden by our socio-cultural emphasis on technoscience as *the* solution to our problems—that is, the search for a 'techno-fix' (Birch et al. 2010). Alternatives do exist, and they frequently reflect very different political-economic thinking than neoliberalism, although sometimes alternatives reflect neoliberal principles and can fit easily within the expectations of the dominant political-economic regime. The tricky question we then have to ask is, are the latter as legitimate as the former? For example, are 'drop-in' biofuels—which I discussed in Chap. 5—a socially, politically, and economically legitimate pathway towards a low-carbon future? As such, sustainability transitions, premised on finding a way to ameliorate the worst effects of climate change, are about trying to balance a set of political-economic principles, or competing principles, with a set of socio-technical pathways; it's reasonable to assume that this balancing act is rarely stable. Hopefully, I've been able to show how bioenergy and especially liquid biofuels represent a socio-technical pathway that is bounded by numerous contentious questions—raising a series of ethical, social, political, economic, *and* environmental issues that are difficult to address one way or another.

SOME THEORETICAL IMPLICATIONS OF MY ARGUMENT

Before I come to the alternative bio-economies out there, I want to turn to the theoretical implications of my argument. As others have noted previously, the way we use a concept like neoliberalism can frequently create a 'boogeyman' (Barnett 2009) or monolithic force (Castree 2006) that seems all encompassing and all powerful, standing against any form of alternative political-economic regime or societal change (Purcell 2016). By unpacking the contradictions within 'neoliberalism studies' in this book (Cahill et al. 2018), my aim has been to analyse critically the claims of scholars who argue that 'nature'—defined as the biophysical processes and systems that sustain life on Earth—is increasingly 'neoliberalized', which is defined as a process of extending or installing markets as the main or only organizing institution in society (see Chap. 2). My purpose in doing this isn't to disparage this research, which is insightful and important work, but rather to think through two assumptions underpinning this approach: first, that material nature is transformed by social, often alien processes (e.g. pricing), and second, then, ontologically speaking, markets are best thought of as disruptors of material natures.

Trying to unpack these two aspects of this 'neoliberal natures' literature, as it has come to be known (e.g. Castree 2008a, b; Backker 2010), led me to examine different ways that nature-economy relations are conceptualized. In particular, I drew on the burgeoning literature on socio-technical or sustainability transitions, which has spawned numerous arguments, case studies, side angles, and so on (e.g. Geels 2002, 2005; Shove and Walker 2007; Coenen et al. 2012; Tyfield 2014; Hansen and Coenen 2015; Calvert et al. 2017), the emerging sub-field of environmental economic geography (e.g. Bridge 2008; Hayter 2008; Heidkamp 2008; Patchell and Hayter 2013; Kedron 2015), and debates on material cultures, politics, and economies (e.g. Pinch and Swedberg 2008; MacKenzie 2009; Mitchell 2011; Birch and Calvert 2015; Becker et al. 2016). I sought to draw these sometimes disparate intellectual threads together in order to develop an analytical perspective that privileged *political-economic materialities.* By this, I mean the co-construction of markets and natures, rather than the assumption that these are inherently in conflict with one another, or that one or the other is always necessarily alien to the other, some sort of aberration to a naturalized working order. My rationale for this was to attempt to avoid the two assumptions highlighted above affecting the neoliberal natures literature.

Using the development of advanced biofuels as a case study helps to illustrate these political-economic materialities and provides me with the means to understand the current state of play with both conventional and advanced biofuels; that is, why are we where we are now. In particular, it has helped me understand four key aspects of the emerging bio-economy. First, I examined the role of visions, narratives, and discourse in constituting societal futures. As outlined in Chap. 4, the imagined futures we present to the world represent a powerful driver of societal change, enrolling social actors, accruing societal resources, and reinforcing a particular 'common sense' about what we *should* and are able to do (Ponte and Birch 2014). Second, I analysed the political-economic materialities of advanced biofuels in terms of what they enable and what they limit. On the one hand, I analysed the form of advanced biofuels enabled by these political-economic materialities in Chap. 5, namely, 'drop-in' biofuels that do not require any wholesale change to prevailing institutions and infrastructures. On the other hand, I analysed the limits to the development of advanced biofuels resulting from these political-economic materialities in Chap. 6, specifically as these limits related to the deployment of market development policies to support an emerging bio-economy. Finally, I sought to bring these contrasting arguments together in Chap. 7, where my intention was to think through the co-construction of markets and natures, rather than conceptualize either as shaping the other. As a result of making these arguments, it's evident to me that the assumption that markets are aberrations of nature—or society—is untenable; markets as we currently understand them cannot exist outwith nature, nor nature as we currently experience it outwith markets.

In my view, then, the key theoretical implications of my argument are deeply normative and/or political. If markets and natures are co-constructed, then markets are always, necessarily material institutions or mechanisms, meaning that they are never pristine, or ideal, or perfect, or any other abstraction that plagues our thinking about them. As Karl Polanyi (1957) and others working in the same *substantivist* tradition note, markets and economies are socio-natural systems; they are not abstract reflections of notions of logical self-interest, rational expectations, or economizing (a la marginal utility maximization). Consequently, when analysing and writing about markets, scholars critical of 'neoliberalism'—and capitalism more generally—could start thinking about how to appropriate markets conceptually, and how to frame them as potentially positive institutions or mechanisms politically, rather than treating them as always negative (Storper 2016). It is time to take back the markets, analytically and politically.

Here, it's also crucial to consider the ways that material markets are constituted in order to think about the ways to reframe and appropriate markets. First, in terms of *what* markets allocate and distribute, and this does not have to mean everything in our societies, or even most things; markets can be used in a very prescribed way to allocate only a certain set of goods and services (e.g. crackers, but not education), delineated by specific or general social and ethical values (e.g. learning for learning sake). Second, in terms of *how* markets allocate and distribute, and this does not have to be through a price mechanism; markets can be used to value resources without monetary valuation (i.e. price). As an example, university researchers (used to be able to) make decisions about what they want to research themselves, and others then determine whether they want to read that research, but the valuation of said research need not ever be price-based (e.g. someone can research ballet their whole lives without it being priced). Third, in terms of *who* allocates and distributes in markets, and this does not have to mean everyone or, conversely, only certain people; markets can be open to everyone, limiting the ability of social groups to create in-group and exclusionary networks (e.g. the 'old boys' club' that dominated universities in the past and still do to a lesser extent).

Turning to these political/normative issues is new for me, as much as it'll be new to many others writing in the literatures I've drawn on here. It's not easy to think about the object of your critique, or scorn, or ire, as a potentially positive thing. That being said, to me at least, it makes political sense to find ways to rehabilitate markets politically, so that they are not associated almost exclusively with centrist or right-wing political ideologies—in which neoliberalism is generally lumped. Of particular importance, in my view, is countering the tendency to depict publics—presumably including ourselves—as subsumed by some form of neoliberal subjectivity or identity, most evidently the work of people like Dardot and Laval (2014) and Wendy Brown (2015). As I've argued elsewhere (Birch 2017), what we end up with in this framing of people as increasingly neoliberalized is the representation of people as *market monsters*, performing a set of market principles both pristine and uncontested. A more politically positive perspective is to think about how material markets reflect our historically specific context and are, therefore, open to political contestation and capture. As a way to illustrate these possibilities, I now want to turn to specific examples relevant to the rest of the discussion in this book about the bio-economy.

Alternative Bio-Economies Out There

Much of the academic and policy literature defines the 'bio-economy' in relation to modern biotechnology and life sciences; for example, the OECD (2006) and White House (2012) specifically associate biotechnology with the bio-economy. McCormick and Kautto (2013) argue that 'biotechnology' is frequently characterized as a 'means' to a broader societal goal, such as transition to a low-carbon future; this is evident in policy agendas around the world, such as the EU's strategy and including the already mentioned OECD and White House strategies. However, there are other definitions that differentiate between the processes at play (e.g. biotechnology) and the resources being used (e.g. biomass), as Staffas et al. (2013) and Birner (2018) outline in their work. Here, however, rather than use 'bio-economy', this resources perspective emphasizes the 'bio-based economy' definition, broadening the object of study from biomass transformed via modern biotechnology to a wider range of agricultural, aquacultural, and ecological goods and services (e.g. food, biodiversity, fish, etc.). As a concept, it's probably best to think of the bio-economy as being the combination of these two perspectives: first, the bio-economy involves the use of modern biotechnology to transform biological materials, and second, it involves the use of biological materials as a substitute for fossil fuel materials (see Birner 2018). As such, the bio-economy can mean many things to many people. Not only does it represent a policy vision that can enrol numerous stakeholders (e.g. farmers, biotechnology firms, environmentalists, etc.), it can also be practised in very different ways—as I outline next.

Organics and the Bio-Economy

One key example of the way that the bio-economy has come to define very different things is the development of the European Commission's (EC) Organics Technology Platform, or TP Organics. It currently sits under the broader 'bio-economy' rubric in the EC's research funding arrangements, called Horizon 2020 (2014–2020), which has a budget close to €80 billion. According to the ETP website, these technology platforms:

> ...develop research and innovation agendas and roadmaps for action at EU and national level to be supported by both private and public funding. They mobilise stakeholders to deliver on agreed priorities and share information across the EU.[1]

Along with colleagues, I've analysed the role of these ETPs when it comes to the bio-economy, including the emergence of alternative understandings of the bio-economy embedded in TP Organics (e.g. Birch et al. 2010; Levidow et al. 2012, 2013). Initially, and unlike most other ETPs, TP Organics was developed outside of the official EC policy framework. Started in 2006 by a number of organics stakeholders, it published a vision statement in 2008 called *Vision for an Organic Food and Farming Research Agenda to 2025*, and then a strategic research agenda (SRA) in 2009 (Niggli et al. 2008; TP Organics 2009; Birch et al. 2010). According to the vision statement:

> Organic agriculture and food production are innovative learning fields for sustainability and are therefore of special interest to European societies ... In order to maintain a leading position in this innovative political and economic field, research activities are crucial. (Niggli et al. 2008: 9)

The SRA identified three key themes it wanted to pursue, including 'empowerment of rural areas', 'eco-functional intensification', and 'food for health and well-being' (Levidow et al. 2012). Critical to developing these research themes was the idea of diversified biomass use and horizontal integration of different values chains (e.g. food and energy), contrasting with the emphasis on vertical integration in other bio-economy ETPs. Eventually, in 2013, TP Organics was officially recognized by the EC as an ETP and integrated into Horizon 2020 (TP Organics 2015).

Within this alternative framing of the bio-economy, which is perhaps particular to the EU, the understanding of the bio-economy obviously reflects the policy vision and framework presented by the EC (e.g. CEC 2012), which emphasizes the goal of moving towards a low-carbon economy through the "sustainable use of renewable biological resources

---

[1] https://ec.europa.eu/research/innovation-union/index.cfm?pg=etp.

for industrial purposes" and "ensuring biodiversity and environmental protection" (Schmid et al. 2012: 47). However, as Schmid et al. (2012) note, much of this EC policy agenda ignores both existing local knowledge and existing agricultural knowledge of farmers and other stakeholders (or 'agroecology'); instead, it emphasizes the role of technoscientific knowledge (or 'life sciences') produced in labs (Levidow et al. 2012). While there might be links between the broad EC definition and the more specific organics definition, there are significant differences. For one, they are both concerned with very different societal challenges: on the one hand, the EC's technoscientific approach focuses on resource inefficiencies; while, on the other hand, the TP Organics' agroecological approach focuses on problems of industrial monocultures (ibid.: 47). The agroecological approach aims to link notions of (high) food quality with public goods through ideas like 'multi-functional agriculture' in which food security and quality are aligned with the maintenance of environmental quality and biodiversity (Schmid et al. 2012). How this plays out on the ground is a good question to ask in the future.

From Bio-Economy to Eco-economy

A number of academics have argued that the bio-economy—and associated concepts—plays a role in naturalizing our relationship to the environment. For example, Pierce (2015) argues that conceptualizing nature as a storehouse of economic value—key to ideas like 'natural capital'—shapes the way people then understand it, while Zwart et al. (2015) argue that the bio-economy is frequently framed as less disruptive of nature, and therefore somehow more 'natural'. Both these arguments illustrate the importance of discourse, which I discussed in Chap. 4, but here it's in terms of the cultural implications of how we understand and then treat the environment. From their perspective, the bio-economy can create the impression that the environment is simply a place from which we can extract natural resources, some of which are more natural than others—so biological resources are better than fossil resources, although both are treated as 'resources' (i.e. having a market value). As a result, the bio-economy can often seem like another example of 'ecological modernization', or an easy technological fix for a range of societal problems (Kitchen and Marsden 2011).

In contrast, people like Kitchen and Marsden (2011; also Marsden 2012, 2017; McDonagh 2015) argue that we need to promote and support an 'eco-economy' paradigm. Here they associate the bio-economy with support for (technoscientific) innovation and (neoliberal) societal change, especially the versions promoted by the OECD and EC. They're concerned with the resulting impacts on local communities and regional economies of this bio-economy vision since it entails "privatised control over both the [biotechnology] techniques and practices" (Kitchen and Marsden 2011: 757; also see Birch et al. 2010). Like the alternative framing of the bio-economy represented by the 'organics' perspective above, Kitchen and Marsden argue we need to promote an eco-economy based on multi-functional geographies that support biodiversity and rural or regional development. So, for example, policies that involve the extraction of biomass from localities and then ship them elsewhere for processing, refining, and use won't facilitate an eco-economy because they ignore the socio-economics of local determination and control within local nature-economy relations (also Marsden 2012). Rather, the eco-economy is strongly embedded in regional economic geographies and has the potential to renew and/or regenerate declining or less-favoured regions without extracting resources, people, or value from localities (e.g. Coenen et al. 2015).

CONCLUSIONS: WHAT DOES ALL THIS MEAN?

I could present more examples of alternative bio-economies, but I think that these two examples illustrate the main issues I want to highlight here. It is perhaps a good time to return to the empirical focus of this book, namely, the development of advanced biofuels in Canada. Looking at this from the perspective of agroecology and eco-economy, it's possible to identify a number of policy implications worth considering.

- *Emerging bio-economies*: it's evident that the bio-economy, especially exemplified by conventional and advanced biofuels, is constituted by dedicated policy action and decisions. As an example of a new 'sector', it represents a good case study of the importance of the 'entrepreneurial state', as theorized by Mariana Mazzucato (2013). By this, I mean that the bio-economy hasn't and won't emerge spontaneously from the 'natural' habit of humans to truck and barter—as 'neoliberal' thinkers like Friedrich Hayek would have it (Birch 2017). A major reason for this is not because we don't have a free market but rather that 'actually existing' markets are always necessarily embedded within socio-material regimes that favour particular sectors (e.g. fossil fuel subsidies). Working out how to support socio-material transitions—which could mean culturally as well as political-economically—is key to promoting pathways to (desired) low-carbon futures. Obviously, if there is no desire for a particular future, then anything goes.

- *Sunk costs and lock-ins*: a major driver and enabler of any socio-material change is, therefore, the configuration of the dominant and prevailing socio-material regime's infrastructure and institutions. In most cases, it's not possible to make radical and wholesale systemic change, whether technological (e.g. replacing petroleum with biofuels infrastructure) or social (e.g. replacing cars with automobility services) (Tyfield 2017). As such, socio-material change tends to follow existing 'sunk costs' in infrastructure and institutions; whether it can engender subsequent systemic change is likely dependent on how that change is legitimated and who ends up in control.

- *Centralized vs. distributed control*: the bio-economy has proven attractive because it can mean something for everyone (Frow et al. 2009)—an environmentalist can get on board, as much as a profit-driven capitalist. Yet, and this is important, it's a socio-material change bound up with questions of legitimation; that is, it revolves around questions of what we are trying to achieve (e.g. ameliorate climate change), who we are trying to benefit (e.g. farmers, rural communities, Global South), and how we are going to go about doing it (e.g. biotechnology, agroecology). Legitimation in this context means enrolling publics and stakeholders in support of a future vision and ensuring that the vision delivers on its promise (Rohracher 2010). Politically, this entails a significant tension between notions of political power and control; that is, it involves contestation between notions of centralized or distributed control over our futures. As just discussed, it means working out how to manage significantly divergent alternative bio-economies (e.g. Schmid et al. 2012).

- *Standardizing sustainability*: finally, the underlying premise of the bio-economy—that it's more sustainable than a fossil-based economy—is not entirely unproblematic. On the one hand, it entails the deployment of market development policies (MDPs)—a process redolent with neoliberal overtones—especially the creation of standards, certification, and accreditation as a way to engender public acceptance (Scarlat and Dallemand 2011). But, on the other hand, this standardization is riven with subjective judgements about what would constitute 'sustainable' bio-economies. The potential for some countries to exert pressure to assert their position is already evident in the way that the EU requires adherence to certain standards (Ponte 2014; Neimark 2016). It's difficult, if not impossible, to ensure that sustainability criteria are embedded in the materiality of biofuels, or other biomaterials. However, these biological products have to embody sustainability in order for it to become a standard reflected in production, distribution, consumption, and disposal processes that ensure 'sustainable' life cycles.

Too often in my academic career I've found myself as the critic, the person deconstructing phenomena in order to identify the problems with one thing or another or the contradictions underlying this or that (see Purcell 2016). My attempt here to think through what it means to say that markets and natures are co-constructed is my way to find some way to be both positive and hopeful. It's an evident truism to say that we make the world, and how we make the world matters. In making the world, however, we're making markets and we're making natures. As such, I've become more and more interested in thinking about how to rehabilitate markets as a social institution, rather than condemn them as a nefarious aberration of some sort of idealized or naturalized social or natural existence. In making the claim that markets entail a (biophysical) materiality—and vice versa—I wanted to think through the implications of doing critique without resorting to concepts like 'neoliberalism'. Doing this has led me to ask questions like, what might markets be if we took back control of them? And, what might we do with them once we have control? Such questions open up the future to a range of possible answers.

REFERENCES

Bakker, K. (2010) The limits of 'neoliberal natures': Debating green neoliberalism, *Progress in Human Geography* 34(6): 715–735.

Barnett, C. (2009) Publics and markets: What's wrong with neoliberalism?, in S. Smith, R. Pain, S. Marston and J. Jones (eds) *The Sage handbook of social geographies*, London: SAGE, pp. 269–297.

Bekker, S., Moss, T. and Naumann, M. (2016) The importance of space: Towards a socio-material and political geography of energy transitions, in L. Gailing and T. Moss (eds) *Conceptualizing Germany's energy transition*, London: Palgrave Macmillan, pp. 93–108.

Berners-Lee, M. and Clark, D. (2013) *The burning question*, London: Profile Books.

Birch, K. (2016) Market vs. contract? The implications of contractual theories of corporate governance to the analysis of neoliberalism, *Ephemera: Theory & Politics in Organization* 16(1): 107–133.

Birch, K. (2017) *A research agenda for neoliberalism*, Cheltenham: Edward Elgar.

Birch, K. and Calvert, K. (2015) Rethinking 'drop-in' biofuels: On the political materialities of bioenergy, *Science and Technology Studies* 28: 52–72.

Birch, K. and Mykhnenko, V. (eds) (2010) *The rise and fall of neoliberalism*, London: Zed Books.

Birch, K., Levidow, L. and Papaioannou, T. (2010) *Sustainable capital?* The neoliberalization of nature and knowledge in the European knowledge-based bioeconomy, *Sustainability* 2(9): 2898–2918.

Birner, R. (2018) Bioeconomy concepts, in I. Lewandowski (eds) *Bioeconomy*, Cham: Springer, pp. 17–38.

Bonneuil, C. and Fressoz, J-B. (2017) *The shock of the anthropocene*, London: Verso.

Bridge, G. (2008) Environmental economic geography: A sympathetic critique, *Geoforum* 39: 76–81.

Bridge, G. (2009) Material worlds: Natural resources, resource geography and the material economy, *Geography Compass* 3(3), 1217–1244.

Brown, W. (2015) *Undoing the demos*, Cambridge, MA: Zone Books.

Cahill, D., Cooper, M., Konings, M. and Primrose, D. (eds) (2018) *The SAGE handbook of neoliberalism*, London: SAGE.

Calvert, K., Kedron, P., Baka, J. and Birch, K. (2017) Geographical perspectives on sociotechnical transitions and emerging bio-economies: Introduction to a special issue, *Technology Analysis & Strategic Management* 29(5): 477–485.

Castree, N. (2006) From neoliberalism to neoliberalisation: Consolations, confusions, and necessary illusions, *Environment and Planning A* 38: 1–6.

Castree, N. (2008a) Neoliberalising nature: The logics of deregulation and reregulation, *Environment and Planning A* 40: 131–152.

Castree, N. (2008b) Neoliberalising nature: Processes, effects, and evaluations, *Environment and Planning A* 40: 153–173.

CEC. (2012) *Innovating for sustainable growth: A bioeconomy for Europe* [COM(2012) 60 Final], Brussels: Commission of the European Communities.

Coenen, L., Benneworth, P. and Truffer, B. (2012) Towards a spatial perspective on sustainability transitions, *Research Policy* 41(6): 968–979.

Coenen, L., Moodysson, J. and Martin, H. (2015) Path renewal in old industrial regions: Possibilities and limitations for regional innovation policy, *Regional Studies* 49(5): 850–865.

Dardot, P. and Laval, C. (2014) *The new way of the world*, London: Verso.

Frow, E., Ingram, D., Powell, W., Steer, D., Vogel, J. and Yearley, S. (2009) The politics of plants, *Food Security* 1(1): 17–23.

Geels, F. (2002) Technological transitions as evolutionary reconfiguration processes: A multi-level perspective and case study, *Research Policy* 31: 1257–1274.

Geels, F. (2005) The dynamics of transitions in socio-technical systems: A multi-level analysis of the transition pathway from horse-drawn carriages to automobiles (1860–1930), *Technology Analysis & Strategic Management* 17(4): 445–476.

Goldstein, J. and Tyfield, D. (2017) Green Keynesianism: Bringing the entrepreneurial state back in (to question)? *Science as Culture*, https://doi.org/10.10 80/09505431.2017.1346598.

Hansen, T. and Coenen, L. (2015) The geography of sustainability transitions: Review, synthesis and reflections on an emergent research field, *Environmental Innovation and Societal Transitions* 17: 92–109.

Hayter, R. (2008) Environmental economic geography, *Geography Compass* 2: 1–20.

Heidkamp, P. (2008) A theoretical framework for a 'spatially conscious' economic analysis of environmental issues, *Geoforum* 39: 62–75.

Jackson, T. (2009) *Prosperity without growth*, London: Earthscan.

Kedron, P. (2015) Environmental governance and shifts in Canadian biofuel production and innovation, *The Professional Geography* 67(3): 385–395.

Kitchen, L. and Marsden, T. (2011) Constructing sustainable communities: A theoretical exploration of the bio-economy and eco-economy paradigms, *Local Environment* 16: 753–769.

Levidow, L., Birch, K. and Papaioannou, T. (2013) Divergent paradigms of European agro-food innovation: The knowledge-based bio-economy (KBBE) as an R&D agenda, *Science, Technology, and Human Values* 38: 94–125.

Levidow, L., Birch, K. and Papaioannou, T. (2012) EU agri-innovation policy: Two contending visions of the knowledge-based bio-economy, *Critical Policy Studies* 6: 40–66.

MacKenzie, D. (2009) *Material markets*, Oxford: Oxford University Press.

Marsden, T. (2012) Third natures? Reconstituting space through place-making strategies for sustainability, *International Journal of the Sociology of Agriculture and Food* 19(2): 257–274.

Marsden, T. (2017) *Agri-food and rural development: Sustainable place-making*, London: Bloomsbury.

Mazzucato, M. (2013) *The entrepreneurial state*, London: Anthem Press.

McCormick, K. and Kautto, N. (2013) The bioeconomy in Europe: An overview, *Sustainability* 5: 2589–2608.

McDonagh, J. (2015) Rural geography III: Do we really have a choice? The bio-economy and future rural pathways, *Progress in Human Geography* 39(5): 658–665.

McKibben B. (2012) Global warming's terrifying new math, *Rolling Stone* 19 July, retrieved from: https://www.rollingstone.com/politics/news/global-warmings-terrifying-new-math-20120719.

McKibben, B. (2016) Recalculating the climate math, *New Republic* 22 September, retrieved from: https://newrepublic.com/article/136987/recalculating-climate-math.

Mirowski, P. (2013) *Never let a serious crisis go to waste*, Cambridge, MA: Harvard University Press.

Mitchell, T. (2011) *Carbon democracy*, London: Verso.

Moore, J.W. (2015) *Capitalism in the web of life*, London: Verso.

Neimark, B. (2016) Biofuel imaginaries: The emerging politics surrounding 'inclusive' private sector development in Madagascar, *Journal of Rural Studies* 45: 146–156.

Niggli, U., Slabe, A., Schmid, O., Halberg, N. and Schluter, M. (2008) *Vision for an organic food and farming research agenda to 2025*, Brussels: IFOAM-EU and FiBL.

OECD. (2006) *The bioeconomy to 2030: Designing a policy agenda*, Paris: Organisation for Economic Co-operation and Development.

Patchell, J. and Hayter, R. (2013) Environmental and evolutionary economic geography: Time for EEG²? *Geografiska Annaler B* 95(2): 111–130.

Pierce, C. (2015) Against neoliberal pedagogies of plants and people: Mapping actor networks of biocapital in learning gardens, *Environmental Education Research* 21(3): 460–477.

Pinch, T. and Swedberg, R. (eds) (2008) *Living in a material world*, Cambridge, MA: MIT Press.

Polanyi, K. (1957) The economy as instituted process, in K. Polanyi, C. Arensberg and H. Pearson (eds) *Trade and market in the early empires*, Illinois: Free Press and Falcon's Wing Press, pp. 239–270.

Ponte, S. (2014) The evolutionary dynamics of biofuel value chains: From unipolar and government-driven to multipolar governance, *Environment and Planning A* 46(2): 353–372.

Ponte, S. and Birch, K. (2014) Introduction: The imaginaries and governance of 'biofueled futures', *Environment and Planning A* 46(2): 271–279.

Purcell, M. (2016) Our new arms, in S. Springer, K. Birch and J. MacLeavy (eds) *The handbook of neoliberalism*, London: Routledge, pp. 613–622.

Rohracher, H. (2010) Biofuels and their publics: The need for differentiated analyses and strategies, *Biofuels* 1(1): 3–5.

Romm, J. (2018) Solar and wind power alone could provide four fifths of US power, *reneweconomy.com.au*, retrieved from: http://reneweconomy.com.au/solar-and-wind-power-alone-could-provide-four-fifths-of-us-power-75235/.

Scarlat, N. and Dallemand, J-F. (2011) Recent developments of biofuels/bioenergy sustainability certification: A global overview, *Energy Policy* 39: 1630–1646.

Schmid, O. et al. (2009) *Technology platform organics: Strategic research agenda*, Brussels: Technology Platform Organics.

Schmid, O., Padel, S. and Levidow, L. (2012) The bio-economy concept and knowledge base in a public goods and farmer perspective, *Bio-based and Applied Economics* 1: 47–63.

Shove, E. and Walker, G. (2007) CAUTION! Transition ahead: Politics, practice, and sustainable transition management, *Environment and Planning A* 39(4): 763–770.

Staffas, L., Gustavsson, M. and McCormick, K. (2013) Strategies and policies for the bioeconomy and bio-based economy: An analysis of official national approaches, *Sustainability* 5: 2751–2769.

Storper, M. (2016) The neo-liberal city as idea and reality, *Territory, Politics, Governance* 4(2): 241–263.

TP Organics. (2015) *TP Organics work programme 2015–2017*, Brussels: IFOAM-EU.

Tyfield, D. (2014) 'King coal is dead! Long live the king!': The paradoxes of coal's resurgence in the emergence of global low-carbon societies, *Theory, Culture & Society* 31(5): 59–81.

Tyfield, D. (2017) *Liberalism 2.0 and the rise of China*, London: Routledge.

White House. (2012) *National bioeconomy blueprint*, Washington, DC: The White House.

Zwart, H., Krabbenborg, L. and Zwier, J. (2015) Is dandelion rubber more natural? Naturalness, biotechnology and the transition towards a bio-based society, *Journal of Agricultural and Environmental Ethics* 28: 313–334.

INDEX

© The Author(s) 2019
K. Birch, *Neoliberal Bio-Economies?*,
https://doi.org/10.1007/978-3-319-91424-4

CPI Antony Rowe
Eastbourne, UK
March 29, 2019

TURKISH DELIGHT

BENJAMIN GROOM

EPS

Prelude

The man smoking a gold-tipped, oval cigarette stood and looked out from the bay window of his library across the small quay to the sparkling waters of the Bosphorous; he exhaled a cloud of blue smoke that drifted lazily in the sunlight; idly he watched the boats and ferries of all descriptions plying their way up and down the great waterway. Of all his houses, this *yali*, or large timber waterfront summerhouse, was his most private and his favourite. Built in the 19th century, during the Empire period, it reflected the Ottoman obsession with European neoclassical architecture of the time, yet managed to combine this style with a more familiar Turkish taste: sliding jalousie screens and lattice shutters divided the bright sun's rays into a chequerboard of shade and light. In the entrance hall behind, stained glass windows cast pools of liquid colour on cool marble floors and the beautiful, intricately woven Persian rugs. Ornate gilded ceilings and cedar doors with mother-of-pearl inlay completed a picture of decadent, yet refined, Middle Eastern opulence.

Footsteps echoed across the floor of the hall and he turned to see a young, dark-haired woman enter the room; she was quite beautiful and, but for the leather, pointed slippers that had signalled her approach, entirely naked. She knelt down and carefully undid the buttons of his fly, carefully extracting his tan penis. He was uncircumcised, and with practiced skill, she ran her agile pink tongue straight into the little hood of his prepuce, teasing it back until it had retracted entirely to uncover his rosy glans. Then she started to suck him into a state of hardness.

She looked up at him enquiringly. He nodded. He was ready.

She bent over the desk that sat in the bay window and thrust her bottom back at him. He looked once more at the busy stretch of water above the girl's back that divided the European from Asian continent, then down at her naked posterior; she was waiting patiently, her two orifices both gleaming with lubricant. Which should it be, he smiled to himself: Europe, her tight, familiar cunt, or Asia, her dark, exotic asshole?

He took a sip of coffee from the cup he held and stubbed his cigarette out in the delicate bone china saucer, setting cup and saucer down on the desk beside where the girl leant, her arms forming two sides of a triangle, her chin resting on folded hands. She seemed so composed, so placid. He removed his pants and his underpants, placing them on a chair nearby then, clothed only in a white shirt, returned his full attention to the girl, and placed a hand on either hip, as if to steady her.

She hoped he would fuck her in the normal way today, in her tight, wet pussy. She wanted him there. When he sodomised her it always hurt – to begin with. He was never gentle. She felt the glans of his warm penis nudge between the full globes of her lower buttocks, sluicing through the runnel of her dripping labia. She hoped he would stop there, where the sensations he was creating caused her to moan softly with pleasure. But the second he drew back, she knew that she must bear down hard and relax her anus. A red cloud of pain enveloped her senses as he pushed in and breached the tight sphincter without mercy, without stopping. She wanted so much to twist away from him, but she dared not. The moan turned into a shriek and her fingers gripped at the desk's edge until they turned white. She gasped for air as if winded. He fucked her ass in total silence, the only sound created by his thighs slapping rhythmically against her upturned, vulnerable buttocks. The minutes passed and slowly the pain receded. She began to breathe normally. She even thrust back at him a little, seeking what little pleasure she could from the hard buffeting that her tender vulva was receiving.

She clenched her anal sphincter as hard as she could to give him even more enjoyment. Suddenly it was all over. He tensed and shuddered, clasping her hips in a bruising grip, as he spurted his seed into the depths of her rectum.

The girl looked back at him, smiling, and he nodded his approval, pulling his softening prick out of her ass. She felt a small trickle of warmth on the inside of her thigh.

"That was good, Akasma, very good. Now, be a dear girl and get me another cup of coffee."

Chapter 1

"One thousand Turkish Lira," said the taxi driver. "In advance."

Lucy Dean hesitated a moment, staring at the lean brown hand extended over the back of the seat. The man was filthy and his face reminded her of the pictures she had seen in books of the men who had tortured the Crusaders such a long time ago. His teeth were blackened, some broken off at the gums, and his left eye seemed to be blind. The young blonde shivered and reminded herself that the porter at the luxury hotel where she was staying had selected this man as reliable. Finally she reached into her purse, took out one crisp bill and paid him. With what seemed to her to be a derisive smile, he took the money and stuffed it into his torn shirt pocket. Then he started the engine and slowly pulled out of the parking lot.

Soon they were weaving in and out of the heavy late evening traffic of Istanbul. He certainly seemed to be a competent driver, she thought, as he skilfully threaded through the human obstacle course in the narrow streets. Suddenly they rose over the crest of a hill and she gasped at the sight. Below she could see almost all of the European side of the ancient city. The moon was nearly full, and combined with the lights of the city, it enabled her to get a panorama a hundred times more impressive than any travel poster she had ever seen. She recognized in the distance the famous Hagia Sophia, the Topkapi museum, the Blue Mosque with its seven minarets pointing at the sky like so many missiles ready to be fired. Beyond that, like a thick serpent curled around the Golden Horn, the Bosporus glistened brightly in the moonlight, separating the main part of the huge city from the darker Asian side. She knew that on the other side where she was going now there were camels and mud houses in some of the small villages, and that the men still wore baggy pants and kerchiefs tied around their heads. That much she had learned from the brochures that she and her new husband had been reading over the past year before finally deciding on Istanbul for their honeymoon.

The picturesque quality of the young boys running through the streets with brass trays balanced on their hands carrying the tiny dipper-like containers of thick black Turkish coffee, the strange babble of voices coming in the windows of the taxi, the whole exotic scene around made her suddenly very sad.

Why, oh why did it have to happen this way, she cried inside her troubled mind.

Lucy had been looking forward to her first night with John for such a long time – she had never dreamed it would turn out this way. It didn't seem possible that only two days ago she was still living in her apartment in Philadelphia waiting for the arrival of the so carefully planned wedding day. Only by the greatest effort of her will could she recall her feelings of that time that seemed so long ago now.

She and her fiancé John Dean had been engaged for over a year before that, and they had gone over every detail a hundred times before the big day. Finally the actual moment of saying the words that would tie them together forever came. Then under the colourful shower of rice and confetti they ducked into the waiting cab which had taken

them directly to the International Airport. John had kissed her passionately all the way there, whispering into her ear that he would prefer to cancel their flight and postpone the Turkish honeymoon until the next day. He had tried his best to persuade her to check into a hotel near the airport, but Lucy was determined to go through with the original plan and let him make love to her for the first time in the exotic setting of a foreign city.

At the airport they had separated long enough to change from their formal clothes to sensible travel suits, shipping her wedding gown and his tuxedo back to her parents' house, and then they began the series of delays and frustrations involved in getting from one country to another.

First there had been the usual three-and-a-half-hour delay in leaving the airport because of heavy air traffic. Then the weather was so bad when they put down for refuelling that there was another six-hour wait. Finally, bleary-eyed and tired, they had arrived in Istanbul the next day, only to find that their reservation had been lost. They had waited almost two hours in the lobby as the agent who had supposedly booked their suite argued and gesticulated like a madman with the hotel manager. Eventually they were shown up to a less luxurious room than the one they had expected, and it was then, after the agent had been given a big tip and the bellhop had smirked his way out of their room, that Lucy had first noticed that her husband John was in a state of angry frustration.

So fatigued that she could barely think straight, Lucy had begun to carefully lay out her new nightgown and bedroom slippers. Then she had collected several toilet articles from her cosmetic case, and, trying her best to give John a girlish alluring smile in spite of her fatigue, she had gone into the bathroom to prepare herself for the moment they had both been waiting for.

The new bride was already very nervous about what was to come, and then, when she had started to close the door, she was half frightened out of her wits by the unexpected pressure on the other side. John had followed without her knowledge and was trying to prevent her from locking herself in the bath.

"Just a minute, Lucy," he had said hoarsely in a voice she hardly recognized. "You've held back long enough."

She had looked into his eyes in surprise at his tone, and the expression she saw there had made her blood run cold with fright. He had looked like a cornered wild animal about to attack, a twisted mask of desperation distorting his face. Before she could answer or react in any way, he had wrapped his fingers around her wrist and jerked her back through the doorway, causing her to drop the things she was carrying.

"Ohhhhh. Don't, John," the young wife had wailed, trying to smile at him playfully. But his grip on her wrist was hurting, and the tears began to well up in her eyes. "Please, John. I have to bathe and fix myself up first," she had complained. "You're hurting me. Let go!"

"In a minute," John had growled at her then. "First get undressed and let me have what I've waited for so long. You've tortured me enough for the past year, and now you're going to come through."

Twisting her wrist violently he had flung her towards the huge double bed and she had staggered to a stop with her back to the foot of the mattress. Her eyes blurring with tears of pain and humiliation at being treated this way by her husband of only a few hours, Lucy had stood meekly where she was, sobbing softly and wishing more than anything else that she were back home in her own apartment. She was so tired and felt so out of place in this situation that she was feeling like a lost child in a big city.

John's harsh voice made her open her eyes again.

"I said strip!" he had yelled, like a stranger she had never seen before. Slowly and reluctantly she had begun to undo the top buttons of her corduroy travel suit while staring blankly at the man she had so recently pledged to spend her whole life with.

Blinded by tears, she had heard the sound of his belt buckle being unfastened and the metallic ripping sound of his pants zipper going down. Then suddenly she had felt his hand under hers on the front of her jacket. At that moment her vision had suddenly cleared again, and she had seen him standing before her completely naked. She had hardly noticed his hands tearing the buttons from her jacket with one forceful downward motion as she caught sight for the first time of his fully erect penis.

It's huge! the astonished young bride had gasped to herself. Her husband's cock stood out from under his hard flat stomach like a majestically rearing cobra, ready to strike into the frightened inner depths of her trembling body.

It had only taken him a moment to tear her jacket and matching skirt from her body, and she sobbed and sucked in her breath as he tangled his fingers in the frilly white lace of her slip. It was the first time she had worn the garment bought especially for this momentous trip, but he had brutally jerked it downward, ripping the flimsy material into shreds before throwing it on the floor.

Then he had grabbed her up in his

muscular arms and crushed her to him, wetly kissing her first on the mouth and then on her neck with fumbling lecherous fervour. She grew suddenly limp in his embrace as though she were about to lose her balance and fall. John had then guided her backwards to the bed while she struggled weakly – and then all her strength ceased to exist. He pushed her back farther, until the edge of the mattress caught her behind the knees and jack-knifed her legs, causing her to fall, arms and legs spread-eagled, her stricken eyes staring glassy-eyed up at the ceiling, her long blonde hair fanned halo-like about her face.

John had bent down over her prostrate form, ripping off her nylon brassiere with one impatient motion, revealing her heaving white breasts to his lust-demented view for the first time since they had known each other. Lucy whimpered forlornly and covered her tear-wet eyes with her arm to block out the sight of John's face. It was distorted by passionate fury into an unrecognizable visage out of a dreadful nightmare.

As she remembered those frantic minutes still fresh and vivid in her memory, the bride of only a day shuddered in the back seat of the taxi and loosed a broken sob from her lips. She had been waiting just as impatiently as John for the first time with him in bed, and he had ruined everything by his insensitive brutal attack. If only he had had the self-control, she was sure it would have been the exciting and beautiful act she had always expected it to be.

But no, she thought, he had to act like a rampant rapist and spoil the whole thing.

Once more her thoughts went back to the events of less than an hour ago.

She had felt him move down over her, his wet lips brushing up her side, and he had fastened his teeth harshly into the nipple of her left breast. She had groaned up at him in pain, attempting to twist away from the sudden sharp torture, but his hands, playing roughly over the firm softness of her thighs and hips, kept her pressed into the mattress. His questing lips roamed wetly over the white palpitating mounds of her breasts, causing a blissful unwilling twitch that descended to her loins below.

She had almost begun to hope that he would wait and turn more gentle and loving in his approach, but then suddenly his hands moved together down to her hips. He had hooked his thumbs in the elastic waistband of her white nylon panties, and, in a single deliberate motion, he had torn them from her recoiling body. Then she was lying before him completely naked, his unfeeling actions making her feel like a cheap whore he had picked up off the streets.

She lay panting under his gaze, frightened and disappointed, her eyes still covered with the back of her arm, listening to him breathing hoarsely as he took in the enticing sight of all her exposed private bodily charms. Then she had heard his voice like that of a stranger standing at a great distance.

"And now you're going to get fucked, you teasing little bitch," he had growled down at her. "Spread your legs and show me your cunt."

Lucy had moaned incoherently at his obscene words, mortified at being spoken to in such a fashion. Despite her fear and weakened resistance, she could not force herself to reopen her legs to let him see her defenceless vagina below. Suddenly she felt his rough hands pressing outward on the insides of her trembling thighs, which finally yielded before the relentless pressure, once more splaying out to the side.

The frightened young bride had heard the hissing intake of breath as he caught sight of her pinkly layered vaginal lips. She knew he could see the tight blonde curls surrounding the narrow pussy slit, and she had reluctantly moved her arm slightly, gazing at him between the twin mounds of her firmly ripe breasts. As he stared at the soft sensitive part down between her widespread thighs, his eyes were glazed in the greed of his desire.

Without knowing exactly why, Lucy

moaned softly and squirmed on the plastic-covered seat of the taxi winding through the streets of Istanbul. Then she caught the look the driver was giving her in the rear view mirror, and she looked out the window, trying to seem unconcerned and unaffected by what she was remembering.

She had seen John's tongue flick wetly over his tightly stretched lips as he stared at her naked exposed genitals, until, with an animal groan that seemed to come from the depths of his soul, he had thrown himself forward on top of her prostrate body. His hands were all over her, grasping desperately at her quivering white breasts, pinching all around the smoothness of her buttocks pinned to the mattress beneath, grappling her firm innocent young body up tighter against his as if he wanted to crush the life out of her. Then his right hand was fumbling down between their bodies, and he had taken the hard throbbing length of his penis between his fingers and guided it forward, using the thick rubbery head to part the warm tender lips of her vagina. She had turned her head to the side on the bed, closing her eyes tightly with a shudder as she felt the blunt tip come into contact with the sensitive edges of her cringing virginal pussy. She gasped and held her breath for what seemed an eternity, lying absolutely still in subjugated helplessness under her new husband, not dating to breathe until she felt his next move.

"Noooo, wait," she had pleaded with him, her voice trembling as she felt the first shocking pressure against the tight elastic opening of her vagina.

He shoved.

"Aaaaaaagh, wait! It's too big!" she moaned in agony as the huge tip slipped through, stretching wide the small, slightly wet cuntal opening until Lucy thought her thighs were splitting apart from the relentless outward pressure.

"God, no, please, you're hurting, you're hurting me!" She had been almost crying as she jerked open her eyes in fear and saw by his face that he had not even heard her. He was lost in the throes of his first real chance to get at her innocent body, and he was completely oblivious to everything else, even her pleading cues.

Suddenly, his demonically twisted face had taken on a contorted expression of sheer raw hunger as he looked down at the lithe blonde-haired woman he seemed to hardly recognize. His young bride was helplessly pinned under his heavy weight, the head of his pulsing young penis buried hotly in the soft curling hair of her cunt. He fell forward then, his heaviness smashing her full, rounded breasts tightly back against her chest. He thrust his hips forward at the same time and his hard thick cock slipped into her pussy like a driving iron rod pushing the moistly resistant flesh of her vaginal walls in rippling waves before it. There was no stopping it until he gave a loud deep-chested groan as his testicles slapped heavily into the quivering cheeks of her tightly clenched buttocks.

"Oooohhhhhh! Ooohhhhhhh!" she wailed beneath him, her legs flailing out to the side as she tried to ease the searing pain of her abruptly ended virginity. She had never been so utterly filled in her life, and his rock-hard, impaling cock felt as though it had torn her vagina into a thousand tiny shreds as he rammed into her without thought of mercy or injury to her tender young insides. And now it lay sunk deep in her belly, filling it to the point of bursting. There was not a single ridge of flesh that she could not feel as the quaking inner flesh of her cunt accommodated the intruding cock, encased in the hot vaginal sheath like a heated dagger planted cruelly into her belly.

Almost without hesitation, the impassioned young husband had then begun to make long sliding strokes in and out of his new wife's tortured cunt, groaning with each inward thrust as though he wanted to pierce her through and through. Then almost as abruptly as it had begun, just at the point when she

had felt her vagina lubricating its passage and easing its smooth strokes, she had felt his cock begin to jerk wildly inside her as he pumped vigorously up into her wide-stretched loins. He mumbled incoherent obscenities above her as his spewing cock exploded, filling her with hotly spurting sperm, finally sinking its ejaculating length in to the hilt and holding it there. It throbbed and quivered like a wounded animal for a long time, while John slowly and painfully ground his pelvis in close circles against hers as she felt his full load of cum emptying into the inner softness of her belly. Finally his hardness was depleted – she felt his penis begin slowly deflating inside her as he collapsed in exhaustion over her nearly lifeless young body.

The ravaged young bride had felt him gradually regain control of his breathing, and then, without warning, he had raised himself up away from her, his now limp penis slipping wetly out of her sore, stretched cunt with a slight sucking sound, a thin trail of hot sticky sperm trailing after it to seep down the smooth interior of her thighs.

Lucy groaned in mortification in the taxi as she remembered what had happened next. John had stood up away from the bed and had looked down at her with nothing but uncaring disdain in his eyes.

"I don't know why I waited so long," he

had spat the words out. "It wasn't even worth the trouble." With that he had hurriedly begun to get dressed, starting for the door as he put his jacket on again. "I'm going out to find myself a good piece of ass," he had hissed brutally at her. "Somebody who knows how to fuck."

And then he was gone, leaving her lying there to feel used and useless in her ravaged nakedness.

She had cried to herself for an hour or more after he had left, and then she had pulled herself weakly from the bed to take the shower he had prevented before.

It was while she stood weeping under the stinging streams of hot water that she had come to her decision to leave the hotel before he could return and subject her to still further humiliation. She had dressed and packed her things back into the suitcase she had so recently emptied.

On a sudden impulse in the lobby she had leafed through the brochures lying on the counter and decided to go to a motel she saw pictured in one of them, close to the historical site of ancient Troy. She had ignored the leering laughter hovering in the eyes of the night porter, finally getting into this taxi in which she now rose through the mysterious looking city toward the ferry to take them to the other side.

Though she had no idea what she would do after she had arrived at the motel, she was certain that she was doing the right thing. If she ever decided to let John know where she was, he would realize that he had to treat her with more respect and tenderness than he had so far. He would have to learn that she had feelings too, and when that happened she was sure she would be able to give him as much pleasure as any of those street whores he had gone out to find.

Meanwhile she would just go right on and enjoy seeing the strange new country she was in – she would try to forget everything that had occurred. Then they might be able to start all over again.

Lucy felt the bump and lurch of the taxi as it passed from the hard asphalt surface onto the ramp of the ferry, and when they had stopped in the depths of the boat, she got out to see if she could find some coffee on one of the upper decks. When she reached the top of the steps leading up from where the cars were, she glanced back at the hazily silhouetted city. Something struck her as strange and then it finally dawned on her. The minarets thrusting up all over the place from the hundreds of mosques were like so many erect penises aiming at the moon. She laughed to herself, remembering that only a few hours before they had seemed like missiles aiming at the moon.

I'll show him who's sexy when we get together again, she whispered softly to herself.

She suddenly heard a light cough near her in the darkness and turned to see a Turk with a dark moustache running his eyes up and down her figure. Than she noticed a lighter patch on his black suit, and looking more closely, she realized with a start what it was.

He had unbuttoned his fly for her benefit! With a crawling sensation of horror running up and down her back, she discovered that he was holding his thick penis in his hand, slowly masturbating as his eyes roamed over the contours of her young body. She cried out in disbelief and hurried back to the taxi below. There she rolled up all the windows and locked the doors from the inside, not daring to move again until the ferry reached the other side and they were speeding off toward the distant motel on the Dardanelles.

Chapter 2

The taxi entering the long driveway into the motel was visible from the night manager's office. Hasad, on duty alone in the nearly deserted motel, was surprised to see the car and wondered who could be arriving at this time of the night. He had been bored by the lack of activity, but now his interest was aroused. He quickly slipped on his maroon blazer and stepped briskly out to meet the car now pulling to a halt outside his office. He could hardly believe his eyes when he saw the young girl sitting all alone in the back seat and his blood began to pound in his temples. Even if this young innocent-looking girl was some rich man's mistress coming to wait for him here, it could prove to be an interesting evening.

When he learned from the driver that she was American and would be alone, it was all he could do to keep from showing the excitement in his face, but he maintained his calm, efficient manner.

"Good evening, Madame," he said in an obsequious voice. "Hasad, at your service." He bowed deeply as he opened the door of the taxi, offering his arm to assist her. Lucy deliberately ignored his gesture and stepped out of the car on her own, casting an aloof glance in his direction, which was not lost on Hasad, although he swallowed his pride and kept smiling at her. There would be plenty of time for him to even the score for her completely unwarranted rejection of his attentions if he had his way.

"Madame wants a cottage to herself? With a view?" He suggested in his senile way. He knew her type very well. As long as he pretended to bow and scrape to her, she would be satisfied.

"You have a vacancy then?" she asked. "They told me in Istanbul that you were hardly ever full. I guess they were right." Secretly the young fugitive wife was wondering why the motel was not jammed with people. Even in the dark she could see that it was a magnificent place to stay. The cabins were scattered about in a pinewood overlooking the water, with a lovely view of the islands scattered in the Dardanelles below. But she knew from her reading previous to the trip that the way to treat foreigners was to keep them off-balance by being haughty and disdainful. It would never do to let them think they had the upper hand, especially if you were an American.

"I suppose that will do," the lovely blonde said as though to nobody in particular, glancing around the office they had now entered as if expecting it to be infested with vermin. "You can get my bags from the taxi."

Without another word, Hasad returned to the taxi and told the driver to put the bags in cabin number fourteen. He waited outside the door until the taxi driver returned ready to leave. Reaching through the window, Hasad gave the one-eyed driver five hundred lira.

"You didn't bring the Madame here, do

you understand?" he whispered in a harsh voice. "She changed her mind, and you took her to the airport."

The driver smiled knowingly and stuffed the money in his pocket. He took a good look at Hasad's face – this was a good thing to remember for the future. Then he slipped the car into gear and pulled away from the lights of the motel.

"If Madame will follow me?" Hasad said when the taxi was gone. He led Lucy with a flashlight along the pathway winding among the isolated cabins until they reached the most remote one of all. The light had been left on inside by the taxi driver and Hasad held the door while she walked in and looked around.

Lucy was still a little dazed from her sudden decision to leave John and from the dangerous trip up the winding road to the motel. She glanced into the bathroom once and wished she were already soaking in the bath.

"All right, the room will be fine," she said at length to Hasad, who was waiting expectantly by the open door. "If you could bring me something warm to drink before I go to bed?"

"Of course, Madame. A hot brandy and milk?" Hasad suggested. With a little special brew of my own, he added to himself.

"Yes. That'll do," Lucy answered. When Hasad didn't close the door and leave her alone, she became suddenly short-tempered and tried to put him in his place with her tone. "Well? What are you waiting for? Please get me the drink now!" She was exasperated. Here she was, tired and frightened and eager to be alone to think things out, and this stupid man only stood there staring at her. It was too much.

"At once, Madame," he said with a little self-deprecatory bow. "If I could please have your passport. Motel regulations require that I fill out certain forms." He smiled at the way she looked flustered and embarrassed under his insistent gaze. Then finally she seemed to understand what it was he wanted, and she dug into her purse searching for her passport. When she handed it to him with an annoyed flick of her waist, she looked coldly up at his face.

"Now will you please bring me the drink so that I can go to bed?" she intoned in a frigid voice. She didn't quite understand how, but this night manager was making her very nervous and uncomfortable and she wished he would leave.

"Thank you, Madame. It will take a moment for the brandy and milk, since the bar is closed. Excuse me," he said, maintaining his cool exterior. He knew that he was making the innocent-looking young woman uncomfortable, and he was enjoying himself. But he knew there was a lot more pleasure to come and hurried back to the office to mix up his special nightcap for young ladies who came to the motel alone.

After hiding the girl's passport, which he had no intention of registering in the guest-book, he took the sleeping powder from his private locker, and after a moment's thought, reached into the inside pocket of his sport jacket to get the tiny lump of hashish he always carried. I might as well give her some pleasant dreams as well, he chuckled to himself as he began to mix the drink. When he had heated the milk on the single burner alcohol stove, he dumped in the strong brandy and sprinkled the two powders on top. He stirred in the whitish sleeping potion and the dark brown hashish carefully; he hummed to himself as he thought of the promising evening he had ahead.

While Hasad was gone, Lucy had quickly changed into a frilly negligee she had brought especially for her wedding night and started running the bath water.

She heard the light knocking at her door, and opened it just wide enough for him to pass the tall warm glass to her. She tried to give him a scathing glance of dismissal, but the way he was staring at her partly revealed figure caught her off guard, and she hurriedly pulled the neck of her negligee close together with her free hand and looked away, closing the door almost in his face. She had never been so affronted in her whole life. His penetrating eyes had bored straight through the negligee Lucy was wearing. She didn't like this night clerk and the way he looked at her. He had stripped her naked with his glances and she knew it wouldn't take much carelessness on her part to provoke him into getting out of line. She had never seen such a raw animal lust in a man's eyes before as when the night manager's locked on the cleavage showing between her firmly ripe breasts. Her hands inadvertently covered them as she trembled in repulsion at the thought of his hands touching her.

God, what a beautiful set of breasts she must have hidden under that gown, Hasad thought to himself after she had closed the door in his face. He had only caught a brief glimpse of the rising swell of creamy-white flesh where her negligee had fallen open, but it was enough to start his blood pumping down in his loins. Without wasting another minute, he let himself quietly into the adjoining storeroom and pushed aside the cover of the tiny peephole he had installed long ago. Putting his eye close to the narrow opening, he could see nearly the whole interior of the bathroom and on beyond through the door to the double bed positioned directly in his line of sight. He caught a slight movement at the edge of the doorway, and he licked his lips in anticipation of what was to come. He noticed with satisfaction that she was taking a deep draught of the drink he had made as she stepped into the bathroom. Then she set the drink down on the edge of the bathtub and hummed softly and sadly to herself as she slowly slipped the negligee from her shoulders.

It was all Hasad could do to keep from groaning out loud at the sight. She was facing over the bathtub directly toward him as she started to remove the flimsy garment, gradually revealing her naked young body to his gaze.

He muttered a faint sigh of lusty appreciation as her fully rounded white breasts burst into the open air of the bathroom. Then – there was nothing left. She stood before his gaze completely naked, like the goddess Venus. He let his eyes run down over the firmly swelling curves of her body, noting with mounting excitement the wide fullness of her hips and the blonde triangle of curls up between her luscious-looking white thighs. He could just make out in her standing position the narrow pink lips of her vagina pouting tightly closed up between her long tapering legs. He felt his stiffening penis jerk up eagerly inside his pants as he watched, and he cupped his excited loins in both hands to better enjoy the pleasant aching sensations beginning to spread through his genitals. Once more he let his vision glide back up over the smooth flat belly to her breasts.

Hasad's mouth watered. He could hardly wait to get his hands and mouth on those tantalizing pink nipples and to twist and churn them into the rock hardness of passion. He smiled to himself as he watched her lift the glass of milk and brandy to her lips again, wincing slightly as she made a face at the slightly odd taste the sleeping powder and hashish gave the drink. He held his breath as she looked into the glass with a puzzled expression on her face, and then relaxed again when she shrugged her bare shoulders slightly and took another long sip. Suddenly she turned her back to him and bent over to take something from the overnight bag she had placed on the floor. Small beads of perspiration broke out on Hasad's upper lip as he stared directly at the fleshy half-moons of her buttocks. He could see the tiny mouth of her tightly puckered anus nestled in the narrow crevice between her ass-cheeks as she leaned over, and the thin pink flanges of her cuntal slit beneath.

The spying Turk groaned and squeezed

his now fully erect cock between his hands through the material of his trousers, wishing that he could ram its throbbing length up into the warm tightness of her cunt without further delay. He felt as though his bloated testicles would explode from the excruciating pressure building up inside, and it was almost with a sense of relief that he watched her turn around again and drop two pellets of bubble-bath into the water. Without waiting for the perfumed capsules to dissolve in the steaming hot water, Lucy stepped over the side of the tub and sat down into the water with a long drawn-out sigh of pleasure. She leaned back with her eyes closed as if about to go to sleep, her fingers curled around the glass perched on the edge of the tub.

Soon her voluptuous young body was completely lost to Hasad's sight as the iridescent bubbles began to lock together around her reclining figure, and Hasad quietly let the peephole cover drop back into place. Very soon he could come back and watch her dry her body and get into bed, but first he had to make two phone calls – very important calls.

* * *

The motel night manager had called Madame Cari many times before. By now she had learned to trust his judgment. Never had he failed to deliver into her hands a girl lovely enough to bring a high price in her very specialized market. The men she supplied with nubile young women were always more than satisfied with the merchandise, even when they were extremely demanding in their tastes. Their insistence on American or English girls in their early twenties, with full but not overly mature figures, blonde hair, and blue eyes had always seemed much too narrow to Madame Cari. She didn't understand why some of the dark-haired, dark-eyed, voluptuously proportioned Turkish provincial girls wouldn't satisfy their requirements, but she had long since stopped trying to talk back to the money these men could afford to pay.

As usual, before going out to view the new prize Hasad was offering her, she chose one of her most alluring clinging silk dresses. She knew that Hasad was lecherous almost beyond belief, and she reasoned that she could always distract him if necessary by using her still well contoured body as a lure to his attention. Perhaps she was no longer fit to be a nightclub belly dancer and stripper at the age of thirty-six, she thought as she looked at herself in the mirror before putting on the tight dress. But she was still considered quite a delicious feast of a woman by most men. And she had never hesitated to use her bodily charms to get anything she might want.

After she was dressed and perfumed to her satisfaction, she opened the wall safe of her modern apartment in the heart of Istanbul and pulled a stack of lira notes from the metal box inside. She counted out seventy-five hundred lira, a little over five hundred American dollars, and hesitated before replacing the lid over the rest of the money. Hasad had seemed very definite on the phone in saying that this girl was something really special. That meant that he was going to up the price considerably if he could possibly get by with it. After a moment's thought, Madame Cari removed another three thousand in cash and placed it with the rest of the money in her purse. He wouldn't dare ask for any more than that, no matter what the girl looked like, she thought to herself. Then she locked the safe once more and covered it with the painting of a naked woman.

She had already sent for her usual driver and he was waiting in the car downstairs when she reached the exit from her building.

"Merhaba," she said in greeting as she slid into the back seat of the long black sedan. As they pulled away from the curb, she noticed with satisfaction that her driver had not forgotten to remove the inside door handles of the roomy back seat. Since that one girl a long time ago had managed to slip out the car door, even though she was half-conscious at the time, he had never neglected to take this precaution. That one had cost him a lot of money, and he wouldn't let it happen again.

She watched the pedestrian traffic of late evening Istanbul with relish as they worked their way toward the water. She was feeling warm and grateful to all the tourist agencies of the world that kept sending her supplies of beautiful young unattached women. She was relaxed and happy as she rode along, secure in the knowledge that she would never again have to be poor – at least not so long as there were young girls out in the world on their own, and lecherous men who were willing to pay to get their sweaty hands on foreign delicacies with fair skin and hair.

* * *

Unknown to Madame Cari, Hasad had at times also dealt with another person in Istanbul, a man named Onan Bey. Onan was rich enough to be able to pay sometimes almost double the price Madame Cari could, provided the girl was his type. His taste seemed to go through phases, however, and Hasad could never be sure he would want the offered victim. Onan always broke in the young women himself before selling them to isolated mountain chieftains, and it was therefore his own taste that mattered, not the demands they might make.

This was the first time that Hasad had called them both for the same girl, but he had two things in mind: First, he thought they might bargain with each other and give him a better price than he might otherwise get; and second, he wanted to be sure to get the delectable young blonde out of the motel before the morning help came on duty. He had had a couple of close calls before, trying to hide girls from view all day, and he knew he would be running a risk in failing to report her in the register this time. It might mean his job if it were found out, and he had too good a set-up where he was to want to lose it.

As the scheming night manager hung up the telephone after the second call, the one to Onan Bey, he glanced at his watch and estimated that it would be at least two hours before either of the interested parties could arrive. He greedily licked his lips and walked hurriedly back toward the cabin where the American beauty should just about be finished with her bath. He hoped that she had also finished drinking the special potion he had given her – then he would be able to do what he wanted with her before Cari and Onan showed up.

Once more lifting the cover from the peephole and looking into the room where

Lucy was in fact now finishing her bath, he leaned forward, his breath coming in short gasps as he saw her carefully towelling her naked body before the mirror. She rubbed the rough-textured material a last time over her full, bursting breasts and down along the insides of her shapely thighs, seeming to shiver slightly when the terry-cloth came into contact with the sensitive pink mouth of her pussy; then she hung the towel on the back of the door. She turned out the light in the bathroom and walked out of Hasad's view into the other room. He could hear her digging in her suitcase for something, and then, before long, she crossed into his line of vision again.

Lucy had put on a filmy nightgown that covered her firm luscious body only down to the tops of her full supple thighs. She was carrying the nearly empty glass Hasad had prepared for her, and she turned it up in the air to drain the last bittersweet dregs into her mouth. Then she sat down on the bed and lay back, her feet facing directly at the peephole where the lustful voyeur was standing. The thin nylon gown snaked its way up over the white flat plane of her belly, exposing the soft blonde silkiness at the junction of her slightly spread legs. Seeing the tempting pink slit up between the sparse curls, he could hardly wait to part the velvety folds with his fingers when he would later sneak into her room.

Hasad's bulging eyes followed the sinuous contours of her hips up over the rising and falling rib cage to the creamy white mounds of her breasts. They were set slightly close together and, through the thin covering of her nightgown, he could see the little buds of her taut pink nipples. Lucy stretched languidly on the bed, revealing even more to his staring eyes the enticing treasure up between her long tapering legs. Then she rolled onto her stomach, her well-rounded buttocks clenching involuntarily as she stretched out one arm to turn off the light in the room. For a moment it was almost pitch black, and then Hasad could see that she had opened the blinds to let in the moonlight. In the soft glow coming through the window he could see her once more roll over onto her back as she prepared her body for sleep. She had one leg stretched straight out, pointing at the foot of the bed, and the other was bent to the side and up, leaving the wide open "vee" up between her thighs fully exposed to his gaze. Then, finally, she seemed to fall asleep.

Hasad waited patiently. His potion had never failed him before. Not only would she sleep so soundly that she would never fully wake up before at least eight hours had passed, but she would have erotic dreams that would make it even more interesting for him.

He would show this proud American bitch a thing or two, this blonde beauty who felt she was so superior that she could dismiss him as so much dirt when he tried to be friendly and helpful and polite. He clenched his fist passionately as he saw her squirm around on the bed in her drowsiness. It was light enough in the room that he could see her slim form stretched sensuously along the length of the bed. Her thighs had fallen apart a little more now and he could easily make out the wispy curls that covered the mound of her lower belly and the tender lips of the gaping slit beneath. He lustfully ran his tongue over his teeth as he fingered the master key in his pocket. He would have to wait a few more minutes. His body was soaked in perspiration now from the very thought of that haughty young bitch squirming in helpless surrender beneath his excited body. He listened to the almost imperceptible ticking of the watch on his wrist, until finally he could stand it no longer.

He returned to the breezeway outside the cabin, carefully tiptoeing to the door of the runaway blonde's room and fitting the key quietly in the door. He opened it slowly, pushing his head into the darkened room, listening for any sounds. There was none, except the soft breathing of the motionless form on the bed. He closed the door silently behind him, locking it to insure that no one would disturb them. He looked intently through the semi-darkness of the room. The head of the sleeping girl was facing up at the ceiling, her eyes shut tight as if in a deep hard sleep, yet she moved slightly from time to time as though troublesome dreams were entering her head from the mysterious other world she had entered.

Hasad moved cat-like around the foot of the bed, not taking his eyes from the reclining figure sprawled back limply on it. She had drawn her bent knee up still higher on the bed till it was even with her hip, the smooth white flesh of the inner thigh gleaming faintly in the dim moonlight. The soft blonde hairs covering the pinkly exposed lips of her vagina were plainly visible to his squinting eyes as they adjusted to the weak light in the room.

He involuntarily drew in his breath with a sharp hissing sound. He had screwed many young, lovely women on this same bed, but never an American and never a woman so innocent and pure as this one, and never one so proud and deliciously alluring. He would never enjoy humiliating anyone so much as this virginal young thing slowly sinking into a heavy drugged sleep.

The thought of helpless mewling grunts of pleasure coming from those sensual lips goaded his thickening penis into rock-hardness. He could feel the hot blood throbbing painfully into its swollen head, where tiny droplets of transparent lubricating fluid had already begun to seep from the sensitively aching tip and smear wetly against his tense belly. Silently he unbuttoned the fly of his trousers and let his heavily pulsing cock burst free of all restraints, slightly easing the pain building in his cum-laden testicles below.

He slowly massaged the thick heavy foreskin back and forth over the throbbing cock-head as he edged his way around the bed toward the now helpless blonde bride lying before him. The drug had done its work well. The hard rod of flesh in his hand was jerking eagerly as he pictured ramming it up between those wide-open thighs, burying its thick head far up inside her haughty little belly.

Hasad removed the maroon blazer he was wearing, and then loosened his belt and let his pants drop heavily to the floor. His hardness stood out fully exposed, his bold menacing erection pointing out over the spread-eagled body on the bed beneath. He slowly unbuttoned his slightly dirty, sweaty shirt and threw it without looking into a chair in the corner of the room. He left his shoes and socks on in case Cari and Onan came and he had to get dressed in a hurry, but he hoped he wouldn't have to leave this delicious young bitch until he had drained them both dry of every ounce of strength in their bodies.

He stood for a moment longer over her motionless body, stroking himself into a rigidity that threatened to explode into streaming white-hot spurts at any moment. He considered for a moment going ahead, for it would give him great pleasure to cum over her sleeping face and sculpted white breasts below. But he decided to wait for a future opportunity.

The lewdly aroused Turk couldn't resist forcing one indignity on her, however, before he fucked her in the cunt. He would be risking an orgasm before he had done everything he wanted to do, but he just had to see those moistly parted sensual lips close around his thick shaft, if only for a moment. He kneeled down on the edge of the bed and turned her head gently toward his throbbing, erect penis.

Then his pulse quickened when the sleeping blonde unknowingly moved her hand to search down over her flat white belly. She pulled her knees up until only her toes touched the mattress. Finally her hand found the opening of her soft, down-covered pussy and she gently fingered the moistening slit up between her thighs. Then she stretched her long tapering legs out flat on the bed, and, arching the muscles of her back, she lifted her tightly clenched buttocks off the bedspread, leaving her out-thrust vaginal furrow nakedly exposed to the mercy of his gaze. Hasad breathed heavily above her as her fingers slowly spread the delicate hair-lined lips until the wet inner folds were pinkly visible to the lewd hunger in his eyes.

As the small mouth-like office eagerly opened to accept the young bride's own probing fingers, Hasad gasped out loud in his rising passion. The drugged woman slowly inserted her extended middle finger up into the soft velvety cuntal folds, at the same time stroking the tiny bud-like clitoris into a quivering hardness. He was hypnotized by the lewdness of the sight as the aroused blonde increased the sensual movements of her body on the bed. She began slipping her finger smoothly in and out between the glistening wet walls of her cunt, her legs suddenly jack-knifing up again, her knees tight against her flattened breasts, her creamy rounded buttocks rising and falling in rhythm to her stroking finger. A gentle purr-like moan of unconscious pleasure floated up from her moistly parted lips – until Hasad could stand it no longer.

Her mouth was still only a scant few inches away from his erectly expanded penis, and he pushed his hips slowly forward toward her sideways-turned face, placing the wet sticky cock-head in the warm crevice formed by her passion-parted lips. He placed one thumb under her nose and the other on her chin, pulling slowly outward until the heavy bulbous cock dropped in between her stretched lips and rested lightly against her teeth, the soft flesh of her mouth forming a heated furrow along its throbbing length. He slowly flexed his hips back and forth, several small droplets of clear cum-fluid seeping from the end of his pulsing glans to mingle with the warm saliva forming inside her burning oral cavern. He could feel the hot air from her slightly flaring nostrils flowing excitingly against his rigid member as she breathed in breathless, heaving gasps.

In a sidelong glance Hasad could see her still shoving her middle finger up into the gradually widening opening of her cunt. God, he could easily shoot his hot stream into her virginal young throat right now just looking at the obscene manipulations. He thought of what it would be like to see her Adam's apple bobbing in her throat as she unwillingly swallowed his fiery load of cum, but he decided against it.

The wildly excited Turk reached one hand down to the hem of the drugged blonde's flimsy negligee, sliding it slowly up her sleek milk-white belly, then on over the gleaming swell of her breasts, until her whole naked torso was exposed to his view. Although he had seen her unclothed body through the peephole, it hadn't excited him nearly as much as having her here now, spread helplessly beneath him where he could touch and fondle her to his heart's content.

With the thumb and forefinger of one hand he reached down and squeezed her soft lips tighter around his hotly throbbing penis, continuing the even stroking motion he was now making into her mouth. The other hand moved over her magnificent ripe breasts, tweaking the little pink nipples alternately between his fingers until he could feel them leaping into eager hardness in response to his touch.

The semi-conscious young woman shifted slightly on the bed beneath him, moaning softly as though she were gradually becoming aware of his presence. He held himself still, grasping his achingly hard cock, afraid for a moment that his drug-potion had not done its work completely. His heavy pulsating shaft slipped lightly from between her parted wet lips, leaving a few almost invisible threads of sticky cock-lubricant trailing behind it.

"Oh, John, my darling husband," she mumbled thickly through the mind-numbing effects of the drugs. "We've waited for so long. Come to me my darling. Please..."

The bride of only a day had become dimly

aware of the maddening waves of sensation coursing through her veins and, in her drugged condition, thought she had seen the shadowy movement of her husband in the room. She was totally unaware that she was half-dreaming, and that her erotic feelings were being caused not by her love for her husband, but by the hashish and sleeping potion in the milk and brandy she had drunk. She knew only that it felt as though a great weight had been lifted from her troubled mind and that now things would be all right between them. John would be considerate of her feelings now and would take her as she had always dreamed he would. She could feel her blood stirring hotly deep within her belly, and she knew she wouldn't fail him this time.

Hasad smiled to himself above her. The mixture had worked its strange magic as it always had before.

The little bitch thinks I'm her husband, he grinned lewdly to himself. He remembered that her passport identified her as a single American girl of twenty-two years, which meant that she must have been married only recently to this John. Perhaps they had never even fucked yet. The thought that she might still be an innocent, untried virgin made his thick expanded cock jerk wildly in his hand, and he stroked the throbbing rod slowly, running his free hand lightly over the full enticing mounds of her breasts and down over her quivering white belly to the mysterious blonde pubic triangle below. Gently he shoved her manipulating hand away from her moistly responding cunt, thrusting his own extended middle finger into the ready and waiting wetness inside.

"Ohhh... darling, darling. Be gentle," the girl droned on beneath him. "I'll make you so happy, John. Please be gentle."

Lucy was dreaming on, her body now becoming alert to the magic caresses turning her tingling body into a hot vessel of passionate desire. Tiny thrilling goose bumps dotted the whiteness of her drug-sensitive skin while hot flashes of sheer unadulterated lust shot up through her widespread loins.

God, how she wanted him! Her body ached and burned to be touched gently and with understanding as he was doing it now.

The new bride pushed from her thoughts the insensitive rape her young husband had subjected her to in the hotel earlier that night. She just wanted to make up for all the time she had denied him and herself the natural ways of merging their bodies into one. She felt as though she wanted him to climb right up inside her, to possess her, to still the thunder and hot lightning that was building up deep, deep inside from the ministrations of his maddening fingertips playing softly over her defenceless nakedness.

Perhaps he would understand her now, understand that she herself had suffered as much as he had while they were waiting – that she had wanted him badly too. Now it was different, she was thinking in her dreamy drug-induced state; they were here together alone, and now they were married and completely free to give each other all that they had to give.

In a mental haze, she passed her tongue slowly over her moist lips, savouring the pungent taste of aroused cock still clinging to them. It did strange things to her – the odour and taste coursing through her entire being like a love potion, lighting little fires in her turgidly erect nipples and causing a throbbing in the nerve ends deep up in her warm throbbing vagina. She writhed and twisted her body on the bed, groaning softly up into the dimly lit room. She could feel tiny dewdrops of moisture gathering down there between her legs as her nakedly palpitating pussy lips began a slow spasmodic contracting, clasping wetly together up between her quivering thighs.

Hasad grinned, his greedy dark eyes feasting lewdly on her unconsciously squirming nakedness.

Gently he pulled his finger from her cunt and moved his knees on the bed, crouching on all fours over her undulating young body, pushing her unresisting milk-white thighs wider apart. Then he crawled up between them, his knees pressed outward against her ankles and his face panting a few inches above the openly spread pinkness of her loins. He passed his tongue hungrily over his lips as his eyes bored down at her gleaming pussy rotating sensuously and expectantly just below his extending tongue. He could see her moist juices glistening in the tiny pink opening of her vagina, slowly seeping and spreading down the full length of the delicious narrow valley of tender young pussy-flesh. Her flowing love-juices trailed on down between the rounded cheeks of her buttocks pressed tightly into the mattress below.

Through half-closed eyes, the drugged young blonde could see her husband's face, dimly visible as he leaned down between her widespread legs. She could feel the flat open palms of his hands pressing outward against her sensitive inner thighs, forcing them even farther apart. Her most secret treasure was offered up to him to use as he wished, and she held her breath as he lowered his face closer and closer and then suddenly – she felt the first searing contact!

The lustful Turk's hot probing tongue lashed out lightly over her quivering, erect clitoris.

"Ohhhhh!" she moaned, shuddering from

the electrifying pressure of tongue on pussy-flesh, her whole body jerking wildly up at him, as his warm moist lips closed hungrily over the entire quaking mound at the base of her belly. His face was lost to her view, buried in the fine blonde fleece, and she closed her eyes tightly again to enjoy the wet tickling kisses he was planting in the tenderly shuddering furrow, his moist tongue flicking serpent-like at the dilated opening of her cunt.

Her hands slipped sensuously down over the quivering swells of her breasts and brushed over the smooth plane of her shivering belly, coming to rest at her loins, on either side of Hasad's avidly sucking lips. Her fingers stroked gently for a moment at the flexing hollows of her inner thighs, tracing the hot flashing sensations coursing through her tender skin, and then she salaciously spread the fleshy outer folds of her pussy farther apart to give his devouring lips fuller access to her hotly throbbing inner crevice.

The night manager's wetly searing tongue shot out again and again, its flicking tip circling the erected clitoris while his lips sucked ravenously, drawing the warm slippery pussy-walls into the steaming hot cavern of his mouth. His sharp teeth nibbled playfully at the pink outer flanges as his tongue darted in and out of the succulent cuntal mouth. Then suddenly his teeth closed lightly around the turgid nub of her clitoris tugging at it as though he wanted to bite it off, his greedily pointed tongue darting repeatedly over its rigid tip.

The defenceless young wife groaned huskily as the hot probing wetness licked against the straining pink bud and she bucked her loins tighter up against his face. Slowly he let her tiny clitoris slip out from between his teeth and began laving his saliva-soaked tongue up and down the length of her desire-moistened pussy-slit. Starting at the lower belly and slithering its way over the elastic-rimmed opening of her wetly clasping vagina, the questing tongue slipped down, down into the crevice of her clenching buttocks, probing in heated search of the tight puckered anus below. Her hips were now grinding impatiently down into the squeaking bed while soft mewling animal sounds escaped from between her clenched teeth.

Hasad worked hungrily up and down her pussy, feeling the curly wet pubic hair brushing teasingly against his cheeks. Never in his wildest dreams had he expected to have such a proud innocent bitch squirming under his tongue and moaning for more. Her groans drove his tongue faster and faster as it worked its way up and down the steaming wet valley of her loins. He wanted her begging for it when he was ready to ram it to her – and she was almost there. He had never seen anything like it, never seen anyone so hot for it, even with the potion. She obviously needed fucking really bad, and she would get exactly what she wanted – this was just the beginning.

The incredulous Turk knew the drugged young bride was too far gone now to fight anything he wanted to do to her and his mind began to form weird erotic pictures of the positions he could put her in and the things he could do at will to her thrashing desire-wracked young body.

He chuckled lewdly as he felt her hands begin desperately clawing at his greasy black head of hair, guiding his face tighter against the palpitating opening of her cunt. Again and again, he ran his tongue into the soft hair-rimmed furrow, flicking at the hungrily clutching cuntal mouth for a moment, and then quickly withdrawing to tease again around the hot throbbing edges of her pussy-slit.

He let her force his mouth down still more firmly over the tight little opening to her wetly clasping vagina. As his eagerly sucking lips rounded and covered the clasping viscous opening, he thrust his tongue deep up into it, bringing a low guttural groan from the vulnerable young woman as her soft warm thighs closed convulsively around the sides of his bobbing face. He could feel the wet cuntal flesh slip moistly around his hotly extended tongue as the walls of her vagina opened and closed in a greedy clutching motion attempting to pull the invading instrument in deeper and deeper. It felt as though the nibbling voracious mouth would pull his tongue out by the roots, devouring it alive.

She snaked her legs around his body and pushed her heels down against his back, pressing his face into the velvety trap until he could barely breathe, his nose smashed tight against the tiny hard clitoris above. He breathed in the sweet pungent scent of her vaginal secretions, now flowing in abundance from her cunt, their aroma exciting his penis to a stiffness he could no longer control. He had to fuck this little bitch soon or he would explode his load of cum all over the mattress!

The only half-conscious Lucy was lost in the fire of the moment. Every muscle in her body was tensed as she strained her hips desperately upward toward that maddening probe between her legs. John had become like a god to her. She had never expected it could be like this, that her husband could bring such glorious sensual feelings to her inexperienced body.

Her love for him incited her further as her up-drawn legs opened and closed around the tormenting head that was gluttonously licking at her flame-seared pussy below. The tendons of her neck stood out as she pulled with all her strength against the tangled hair of his bobbing head.

"Ohhhhh! Ohhh, gooood!" she moaned, splaying her legs out wider and wider to give him easier access to her wildly throbbing loins.

The lust-crazed Turk could stand it no longer. He grabbed the hallucinating bride's flailing legs behind the knees and shoved them roughly back against her shoulders, slithering up over her sweat-soaked body at the same time. His rigidly jerking cock brushed against her wet prickling pubic hair as he planted his hands on either side of her shoulders, her ankles locking tightly behind his neck. He could look down between their bodies and see her upturned ass-cheeks and expanded cuntal slit, visibly fluttering just inches from his painfully throbbing cock.

Hasad grinned obscenely above her as he edged his quivering penis slowly forward, watching its broad tip nuzzle into the pink moist flower of her eagerly waiting cunt. He stopped when the tip was firmly set in place, ready to slip forward and impale her at the first lunge he would make with his hips. He could feel her straining upward to take it in her, her hotly palpitating cuntal flanges writhing against him, sending waves of unbearable pleasure all the way from his bloated glans down the rigid cock-shaft into his aching balls which still hung heavily below, but which were beginning to tighten in their scrotal bag. He let her beg with her cunt a moment longer, poised above her like a bird of prey just waiting for the right instant to swoop down and spear her – then he drew back slightly for the first thrusting drive up into her quaking white belly.

At that moment, he heard the unmistakable sound of tires crunching in the gravel of the driveway outside the cabins.

"Damn!" he cursed out loud.

He had put off the exquisite moment he had been waiting for one minute too long. It didn't seem possible that so much time had passed, but it must have. He hovered above her a little longer, considering going ahead and shoving forward into the squirming hot cunt clutching at his pulsating shaft – but he wanted to carry it to full humiliating completion when he did.

Cursing again under his breath as he gazed from above down at the uncomprehending young wife, he disentangled himself from her encircling arms and legs and climbed hurriedly down off the bed.

"Don't worry, little hot bitch. I'll fuck you later till you can't walk," he growled down at her undulating, unsatiated body. It only took him a split second to slip his clothes back on again, and then he let himself out of the room, taking care to lock the door securely from the outside so that the desperate blonde wife would not be able to get away. He left the key in the lock, hoping to return quickly and finish what he had started.

Lucy thrashed and moaned on the bed, unable to comprehend what was happening. Had she somehow made John angry with her again? Didn't he understand that this time she was ready for whatever he wanted to do to her? She moaned out her frustration, clamping her thighs tightly together. She was willing to give him any kind of pleasure, if only he would put his exciting hardness into her now and drown the gnawing fire that raged out of control in her vagina.

The hopelessly confused young wife opened her drug-fogged eyes and glanced searchingly around the dim moonlit room, groaning again in unsatisfied passion, when she found no sign of her husband. Finally, in desperation, she thrust her hands back down into the steaming hotness of her loins, slipping her fingers lewdly along the moist pussy-slit. It relieved the aching urgency a little, and she began to stroke her extended fingers back and forth along the quivering vaginal mouth, driving her excitement to even greater heights. She found her hard-nubbed clitoris with the fingers of one hand and massaged the throbbing little bud in frenzied need, the other hand becoming soaking wet as it plunged into her moistly expanded cunt and established a wild fucking rhythm. She groaned again and again, in frustrated memory of the hard fleshy rod that had been so near to quenching the fires of love "John" had ignited inside her.

Chapter 3

Outside in the cool night air, Hasad raced up the narrow path to head off whoever had arrived at the motel office. As soon as he had rounded the last row of bushes surrounding the cabins, he could see it was Madame Cari. She met him as he ran panting up to the tiny porch outside.

"Good evening, Hasad," she greeted him with a winning smile. "Where's the fabulous little prize you promised me?" She could tell that he was excited to a fever pitch, and automatically her eyes dropped to the swollen bulge in his trousers. The slimy little bastard has been up to something already, she thought to herself. If she really was as lovely and appealing as he had indicated to her on the phone, she hoped he hadn't marred or disfigured her body.

"As usual, she's in cabin number fourteen,"

Hasad replied, not missing the interested glance Cari had thrown at the still throbbing cock in his trousers. You can have some of this too, he thought obscenely. He had long wanted to fuck Madame Cari. He knew she had been one of the most famous courtesans in Turkey when she was a few years younger, and he would like to give her a sample of what a real man could make her feel. He had an idea tonight might be the night, and his cock jerked violently at the thought. He let his eyes wander greedily over the swelling breasts filling the front of her tight dress. He looked on down to the gracefully flaring hips and full rounded thighs encased in the clinging material below.

His lewdly forward glances were not lost on Cari, and she involuntarily shuddered at the thought that he might run his hands over her body if he ever had the chance. She had never been a woman to turn down a good fuck, but this Hasad was just too repulsive. Still, if it had to be done someday, she reflected, she supposed she could do it. He did seem to have a monstrously huge cock. She pushed these unnecessary thoughts to the back of her mind and got down to business again.

"Let's see this fabulous goddess you've found," she said with obvious disdain for Hasad's cock-sure attitude. "I haven't got much time, as you know. My services are in constant demand in Istanbul. I may be losing thousands of Lira just by being here."

"Of course," Hasad answered in a mocking voice. You phoney bitch, he added to himself. You know that most of what you make depends on the young girls that are delivered to you. Without another word he lead the way toward the cabin. Let her see what I've got for her first, and then we'll see what I can get out of this.

They moved silently to number fourteen and he took her into the storeroom from which he had watched Lucy earlier. Sliding the peephole cover to one side, he looked in first and, deciding that the view was not clear enough, took the chance of turning on the light. He had an exterior switch for this purpose, and without hesitation he flipped it on. Lucy was so far gone now in her dreaming sexual fantasy that he hoped she wouldn't notice.

"There she is," he told Cari, a wide grin on his face as he stood aside to let her see the blonde treasure that would be up for sale to the highest bidder as soon as he was finished with her.

Cari put her face to the peephole and looked into the room. She gasped as she saw the innocent young blonde on the bed writhing in the throes of masturbation. She could see her long tapering legs flung wildly up into the air, her toes pointing at the ceiling, her grinding buttocks beneath gleaming in naked exposure. Between the girl's out-flung thighs, she could see the white half-moons of her perfectly rounded breasts, and above them the lovely face surrounded by a careless halo of blonde hair. The girl's lips were parted in heated passion, working crazily as she tried to bring herself to her obviously much-needed completion. Then the spying Turkish woman let her eyes drop to the openly spread vagina beneath, the fleshy pink folds cruelly distended to make room for the girl's fingers stroking in and out of the clasping narrow passage.

Then the light went out, plunging the room into semi-darkness again.

"Well?" Hasad demanded with a triumphant tone in his voice.

"You've drugged her!" Madame Cari accused him. She didn't want to lose her advantage and let him know that she wanted to get this girl more than she had ever wanted any of them before. The shrewd businesswoman could make a fortune this time if she played her cards right, but she had to keep his price down if she could.

"What difference does that make," he snarled back at her. She was aware of the effect the brief glimpse of the girl had had on Cari. He felt secure that this time he could get whatever he wanted, and Cari wasn't going to bargain him down. In fact, he told himself, I'm going to get more than just money this time.

The sight of the girl finger-fucking herself had aroused him almost beyond endurance all over again, and, if he couldn't get back in the room to satisfy himself on her, then he would fuck this Cari as part of the deal he would make with her. And he was sure she would agree to it, especially as soon as she knew that Onan Bey was on his way out from Istanbul too.

"All right, let's go where we can talk," Cari said. She could see that Hasad felt he had a distinct advantage in this negotiation, and she was anxious to find out what it was. She steeled herself inwardly for the worst. If he insisted on fucking the shrewd Cari herself, she would do it – the girl was worth even that. And besides, she reflected, the obscene view she had enjoyed of the young girl trying to satisfy herself had set her own blood to boiling, and she could feel the beginnings of desire tingling up inside the warm recesses of her excited cunt.

Hasad led her to an adjoining cabin and unlocked the door. Once inside the room, he offered her a drink, which she refused, and then got down to talking business.

"First, I have to tell you that Onan Bey has been notified of our little blonde bitch," he grinned at her, watching with secret glee as her features fell into a pained expression of betrayal.

"What! You lousy bastard!" she screamed at him before she could catch herself. Immediately she regretted her words as his eyes narrowed to angry slits in his face. "Forgive me," she added in a vain attempt to mollify him. "But you know that it's impossible for me to compete with Onan. He's one of the richest men in Istanbul."

"Exactly," Hasad gloated. "Therefore it should be obvious that you have only one advantage over him that might swing the deal in your favour." He ran his eyes hungrily over her stocking-encased thighs, exposed almost up to her crotch as she sat on the edge of the bed opposite him.

"I don't understand," she said, smiling weakly as though completely innocent of what he was thinking.

"I think you do, Madame Cari," he growled. "For you, I might sell this American for only slightly more than the regular price, but only if you please me as part of the bargain." He stood up and glared at her with an obscene leer on his face, waiting for her answer as though he didn't really care what it was.

Cari stared at him for a long time, unwilling to admit that he had her in a position where she had no more power over him. For the first time, he had gained the upper hand with her, and she didn't like the feeling at all. Finally, thinking that time was more important than ever now that she knew Onan was on his way, she sighed heavily and stood up. She slowly began to remove her clothes, lifting her sheath-like dress over her head and throwing it carelessly onto the chair beside the bed.

Then she stood before the gloating Turk in nothing but her black nylon bikini panties and brassiere, feeling more degraded and humiliated than ever before in her life – even worse than that day so long ago when she has been raped by soldiers during the civil unrest in her native Anatolia. She felt dirty and used under Hasad's relentless grinning gaze, and she decided to pursue the last part of the bargain before he had forced her to submit to his depraved demands and could then deny her the prize and laugh in her face. She knew him well enough to be sure that he would stand up to whatever he had agreed to in advance-especially since she was one of his best customers. That wealthy Onan would only come through when the girl was as perfect as the blonde locked in the cabin next door; Hasad was smart enough to know that he couldn't turn his back on Cari completely, even though this time he could make things tough for her.

"All right," she sighed heavily. "I'll do

whatever you want. But first, let's come to an agreement about the girl."

Hasad saw what she was up to, but it didn't matter. He was getting something he had always wanted and had never been in the position to get until now. He spoke with a shrewd cast in his eyes.

"Five hundred dollars for the girl, in addition to getting my way with you – and then fucking her before you take her away," he said with calculated alacrity.

"That's robbery, you bastard," Cari replied without being able to stop herself, instantly biting her tongue at losing her cool again.

"A girl like that is worth at least two thousand dollars to Onan Bey," he replied coolly, flinching at her name-calling. I'll get even with you in a minute, bitch, he mused. I'll show you who's a bastard. Hasad watched her expression change as he made this statement. He knew by the sudden frustration that crossed her face that he could almost name his own price now. She knew that she was in no position to argue with him for too long since speed was of the very essence if the plans were to be completed before Onan arrived.

"All right, I'll accept those terms, but you'll have to come to Istanbul to get at the girl. There's no time here tonight," Cari finally answered. I'm not taking a chance on

Onan getting here and spiriting that girl off before I have a chance to get her away, she was thinking.

"Then strip down, and let's seal the bargain." Hasad countered. He knew exactly what she was thinking and counted on it to get her to do all the things he wanted. Finally he had found a way to equalize himself with her. He knew there was a chance of losing Onan as a future customer, but that was a risk he would take. Right now all he wanted to do was fuck her senseless and try to get rid of her before Onan showed up. He could always say that the American girl had escaped before he could do anything about it, if it came to that.

Cari detected the bold, more masterful change in his voice. It hinted at a certain unmistakable viciousness that suddenly curdled her stomach. God knew what depraved acts he would force her into before this ordeal was over. She wished she could call on her driver for help, but now there was no way out.

Hasad saw her reluctance and once again offered her the drink of straight brandy she had refused before.

"I think Madame Cari will need this. The agreement will not be valid without total cooperation. Do you not agree?" He smiled triumphantly when she numbly nodded in assent. Then she grabbed the drink he was offering and poured the burning liquid down her throat.

She felt as though she would scream in revulsion if this swine touched her, but it was coming and she had to deaden her senses. Things had gone too far to turn back now and she could not afford to lose this young American – it would ease her heavy expenses that had been getting higher and higher as she had been forced to pay larger and larger bribes to the authorities to keep herself protected.

Hasad knew the moment she nodded her head that the battle was won. He was going to fuck this high-class bitch who had once been in demand by the most important men in Turkey, and there was nothing she could do to stop him. He smiled lewdly as he stood in front of her unbuttoning his pants and then letting them drop to the floor. His throbbing hard penis stood out from his body pointing straight at her. It looked like a giant wooden shaft growing up through the black underbrush of his thick pubic hair. With one hand he stroked the foreskin back and forth over the expanding cock-head that grew in size each time it disappeared and reappeared through the thick flesh covering it. He watched the loathing in her face as her eyes remained involuntarily locked on his cock, his excitement flaring as he saw the hopeless fear rising in her eyes. This would be more fun than the American. This one would be conscious of the things he was doing to her!

"Strip!" he hissed at her again. "Or do I have to do it for you?"

Hasad stepped toward her, his mouth open, his eyes drinking in the long tapering roundness of her silk-covered legs, the bursting fullness of her brassiere, the whiteness of her flat smooth belly above the top of her sheer black nylon panties. His gaze made her feel as though she were back at the beginning again, laying her body down for whatever filthy sailor was willing to pay for her services. She thought of all the years she had spent working her way up in the world until now she lived almost like a princess, realizing they were being swept away in this one brief encounter.

"Don't touch me, you beast! Don't touch me!" she shouted without thinking.

"It's too late, Madame," he intoned with contempt in his voice. He grabbed her shoulders with his hands, the strong sinewy fingers digging harshly into her skin. "We have our agreement about the American girl to consider."

"I don't care! I'll find someone else. Let Onan have her!"

He loomed above her, his eyes devoid of pity, shining coldly into hers, his cruel and unyielding lust boring into the very depths of her soul. The pressure of his hands grasping her shoulders permitted no escape from his contemptuous gaze.

"No, no. I can't do it," she whimpered.

Her pleas fell on unhearing ears as his arms enveloped her, his lips crushing down against her, his thick rigid cock below pressing hard into her softly yielding belly. His tongue snaked its way out between his sharp teeth into the resisting wetness of her mouth. She struggled to free herself, but fear and the strong brandy had drained her of the strength to fight. His slim muscular body glued itself tightly to hers, arms and legs winding about her like a giant octopus from which there was no escape.

"Please, please don't," she groaned into the searing mouth locked over hers, the savage rape she had suffered in that earlier time whirling dizzily through her mind. The room seemed to spin as he pushed her backwards toward the bed, sending her sprawling flat on her back, his body descending on top of her to pin her tightly to the jiggling bed. She pressed her thighs together, attempting to hold back the squirming body trying to lodge itself between them. His throbbing cock was trapped there, forcing itself lewdly up and down against the thin nylon strip of the panties covering her cringing vagina below. Cari could feel the hard thick rod sliding against the sensitive inner hollows of her struggling thighs as his head pressed down forcefully against hers, then suddenly dropped, and she felt the sharp excruciating pain of his teeth biting savagely into the taut muscles of her straining neck. She kicked out automatically with her long smooth legs, attempting to dislodge him from above her.

His body sank triumphantly down between her protesting thighs as they splayed obscenely open, and the broad pulsing tip of his cock-shaft pressed demandingly against the protective nylon band of her panties. His pelvis began a deliberate grinding motion against her upturned loins, rubbing the sheer material into the dampening slit of her panty-covered pussy. His hips undulated slowly, slowly, watching the changing expressions on the face of the beautiful woman trapped beneath him, knowing she couldn't stand this torment for long. He gloated to himself that he would soon ignite her body to his lewd probings, and then she would give him what he had wanted since he had first started working with her – her complete submission to his will.

Cari could feel it happening. Her long tense form was pressed tightly into the mattress, so that she could do nothing to escape the hotly palpitating rod forcing itself against the trembling moist furrow between her legs. But her unconscious mind was fighting the torture of the teasingly stroking shaft, fighting against surrender to it.

"Noooo, nooo," she mumbled incoherently, even as her hips began to move involuntarily in hardly perceptible little circles that betrayed her determination to resist.

Hasad felt her unwilling response as the thighs that had been pressed tightly together in vain resistance to his prodding cock suddenly fell loosely apart. Her heels hooked behind his knees and, with a low animal-like groan, her arms snaked around his neck pulling his mouth down to press wetly over hers. She sucked his tongue voraciously into her mouth, soft mewling sounds escaping from her greedily sucking lips. She ground her suddenly hungry loins up toward his rock-hard cock, attempting to draw its thick length through the flimsy material of the panties still separating their flesh.

Hasad reached down between their bodies, savagely ripping away the narrow crotch-band with one motion of his hand and then guiding the end of his violently throbbing cock up between the now unprotected ridges of her cuntal lips. He could feel the curly tickling pubic hair parting before his unimpeded onslaught. The thick rubbery tip of his penis momentarily met resistance at the entrance to the hot moist passage and then he felt the elastic mouth suddenly give, allowing his turgid hardness to slip wetly inside. The lusty Turk thrust forward with all his strength and fury, his massively skewering cock slithering deep, deep inside her wet velvety cunt with a heedless shoving force that brought a muffled scream from Cari's contorted face. He didn't relent in his searing plunge up into her quivering unprepared vagina until he felt his balls slap heavily against her upturned ass-cheeks and he held it there, savouring the hotly squirming cunt palpitating around his deeply imbedded cock.

Abruptly the impaled woman no longer wanted to resist as she felt her vagina becoming wet and open around the heavy throbbing rod. She started screwing her cunt up and down on it, pinning her legs back against her shoulders, wanting to take it all to the hilt, until it bored deep into her quaking belly, until it fucked all the way up into her throat. She no longer had any pride left. In her mind, she had been thrown all the way back to her first time, when she had been fucked against her will over and over again until, in order to save her sanity, she had been forced to learn to enjoy it.

The delighted night manager wasn't sure what had happened, but he knew that he had somehow tapped into the central whore that lives inside every woman. He braced himself on his knees and elbows above the wildly thrashing body, letting the hungrily clasping cunt suck his flesh until he retaliated by bucking forward on her upward stroke several times, driving the hotly expanding cock-head right up to the sensitive wall of her cervix.

"Ohhh, ohhhhh," she groaned out of control as his whole invading hardness fucked into her, the power of his thrusts driving her heaving buttocks pitilessly down into the mattress below.

Wasting no time now that he had her responding with every inch of her wantonly rocking body, Hasad reached back underneath her grinding ass-cheeks and slipped his extended middle finger along the widely spread crevice between their rounded half-moons. Finally he found the tiny puckered anus and thrust his finger all the way up to the second knuckle in the tightly clenched little opening.

"Aaagggh, aaagggh!" she grunted in pain at the unexpected intrusion into her cringing anus. Her feet jerked erotically in the air above his back, her toes curling spasmodically in her sheer stockings. He had hurt her for a moment, but she liked the sensations now coursing through every nerve in her ravaged body.

Through the thin wall of moist flesh separating his finger inside her anus from his pummelling cock slipping wetly in and out of her hotly clasping pussy, Hasad could feel the ridges on the bottom of his thick bloated penis rippling through the warm waves of flesh.

Cari began unconsciously moaning all the obscene words she could think of between her panting gasps as she bucked and twisted under his weight.

"Fuck me, you bastard! Fuck me! Ram it in me good!" she groaned over and over again.

Hasad grinned lewdly down at her lust-distorted face as he rammed his pelvis forward with a vengeance. Each time she gyrated her loins down over his smoothly sliding hardness, he thrust the finger in her anus upward until it was fully submerged in her wildly jerking rectal passage, and then, as she bucked back up away from his skewering finger, he drove his cock still harder into the wetly sucking inner recesses of her cunt.

He had her hopelessly impaled between the twisting finger and thrusting cock, and he was enjoying her wildly thrashing movements beneath him more than he had ever enjoyed fucking a woman before. The lust juices from her dilating cunt ran down over his hand, lubricating wetly the finger now sunk all the way to his fist in her rotating anus.

The wildly responding woman strained back under him, arching her loins against his grinding pelvis and twitching back down onto his wiggling finger. She moaned incessantly, her head flailing from side to side on the crumpled bedspread, her body a mass of electric tingles that shot through it half in pain and half in pleasure.

"Uuuuggghh!" she grunted as he began buffeting her harder in lewd rhythm between his hand and the growing plunging cock. He could feel its thickness expanding more with each thrust he made up into the wet hot passage, its relentless lust fed by the very hopelessness of the unwilling woman now squirming incoherently beneath him.

Cari could feel the stone-hard shaft thickening incitingly inside her ever-moistening vagina. She was given no relief from the hotly building crest of pleasure climbing, climbing deep in her trembling belly.

"Harder, harder, fuck me harder," she chanted in time to his long hard strokes. Huge warm waves of delicious feeling raced through her entire body, until she felt like it was an expanding balloon of joy, filling, filling, until she was ready to burst. She sucked wildly on the tongue flicking into her mouth, wanting her every bodily orifice to be filled. This was all that was real, there was nothing else, and then suddenly, with a deep-throated groan, she felt great floods of her hot cum-juices throbbing from the rippling walls of her cunt, streaming out in endless gushes around the hard fleshy rod slipping faster and faster into her. She seemed to be cumming forever, the fluids gushing out onto his heavy balls beneath and over his hand skewering into the crevice of her ass-cheeks.

She gave one long low moan, splaying her legs high into the air and as wide apart as they would go to give the still rapidly plunging cock even greater access. She thrust her loins up at him with superhuman force, screwing herself up hungrily on the thick pumping shaft. Juices flowed still from her quivering vagina as her nostrils flared in passion – then one long last gasp of breath rasped up out of her lungs as though she had been struck in the stomach. She collapsed under him, her body shivering uncontrollably from the lingering sensations following her down from the peak of wanton lust she had reached.

Hasad felt her surging climax as he drove his cock deeper inside her while her legs splayed out and waved on either side of his body. He could feel the hotly jetting stream begin in his cum-inflated balls and rush down the length of his pulsating thick cock-shaft, spewing wildly out into the depths of her clasping womb. He was filling her completely, his hot sperm overflowing with her own love juices out of the curl-lined lips of her spasmodically contracting cunt. He gave one last low gasp as with a jerk he emptied the last of his load of semen into her still quivering belly, and then he too collapsed like a broken puppet over her limply spread-eagled body.

They lay still, a loose tangle of arms and intertwined legs, their breathing gradually returning to normal during the long moment of quiet. Then Hasad lifted himself from the unmoving body of the satiated woman, his deflated cock slipping with a lewd sucking noise out of the cum-soaked furrow between her openly splayed legs.

"It's a pleasure to do business with you, Madame Cari," he tormented her, smiling obscenely down at her still wantonly sprawling body. Then he stood up and dressed quickly, running his eyes over the length of her legs, now spread with such careless abandon, the tatters of her torn panties bunched up around her slowly heaving belly. Her stockinged knees were bent outward away from each other on the bed and he could see the sticky pool of their mingled cum-juices forming beneath her buttocks. Her mouth hung slackly open in exhaustion, and her hair was matted on her forehead and lay in tangles under her face turned to one side.

He directed a final triumphant grin at her. She had been really fucked and they both knew it. Weakly she raised herself from the bed to dress and follow him to get the American girl she had paid so dearly for.

It seemed to her that he had hardly gone out the door when she heard his footsteps returning at a fast running pace.

Hurriedly she slipped her dress down over her head, wondering what had gone wrong.

Chapter 4

Onan Bey was a dark-haired, dark-eyed man in his late thirties. He had been wealthy for most of his adult life, having made his money in all kinds of illegal endeavours, and he liked to live well. This showed in his elegant tailor-made clothes and even his love of good food had not spoiled his fit, muscular physique. One of his prime enjoyments in life was to find new innocent girls, preferably blondes from other countries, and to gradually lead them into a life of degradation and humiliation.

His usual way was to get his hands on them first and use their bodies in any way that appealed to him, before finally auctioning them off to the highest bidder. There were plenty of men with money to enable Onan to live well on what he earned from this alone. But the money was not the main thing. The principal joy was the humiliation itself.

When he had received the late phone call from Hasad at the motel, he had ordered his car to be brought around. Hasad sometimes made mistakes in choosing girls for him, but not often. Now as Onan sat in the black limousine on his way to the motel to view his prospective victim, he smiled to himself in anticipation. The reflecting headlights of passing cars caused the gold teeth in his mouth to flash when he grinned. It made his face seem much more sinister than it usually did, and the evil glint in his eyes as he thought about the new girl gave him an even more depraved look.

He lit up a cigarette, the gold-tipped brand he had specially made for him, and contentedly sucked in its rich, aromatic smoke as he sat comfortably back in the seat. In his hand was a glass of milky white Turkish raki, and he sipped occasionally at the licorice-like drink, absentmindedly listening to the record he had put on the phonograph in the back of the driver's seat. He luxuriated in all these expensive toys, feeling warm and complacent in the knowledge that he was on his way to purchase still another one – one of flesh and blood, his favourite now that all the other pleasures of life had begun to bore him a little.

He saw the entrance to the motel ahead, and leaned forward as the driver turned in. He saw the car parked outside the office and, without a moment's hesitation, he spoke into the microphone connecting him to the driver.

"Stop the car, Mahmet," he ordered in a cold voice. "And turn out the lights."

Silently Onan opened the car door and stepped onto the gravel driveway, standing up and heading immediately for the grass beyond. His highly polished black leather shoes glistened in the moonlight, betraying no sound as he made his way carefully toward the parked car ahead. Suddenly he stepped into the shadow of the nearby shrubbery. A man came out of the office and got into the car. He looked as if he were waiting for someone, lighting a cigarette and taking a few puffs before hanging his hand out the window in a relaxed way.

Onan approached the car from the side and waited at a distance of a few yards to make sure his first impression was correct. Then the man turned his head to the side so that the light from the office shone directly on it, and Onan knew. He cursed under his breath.

It was Madame Cari's driver!

That meant that Hasad had called her about this girl too. Onan was very sensitive about this kind of thing. First of all, he never admitted to anyone, not even Madame Cari, that he was in the white slave trade, and second, he never liked to be forced to compete with anyone else for a prospective girl. And Madame Cari was one of the most notorious operators of all the Istanbul houses of prostitution. She would only too gladly spread the word if she knew that he had been dealing with this low-life Hasad.

Onan hesitated a moment in the dark, and then on sudden impulse, he darted unseen across the driveway and made his way toward cabin number fourteen. At least he could observe what this pig Hasad had thought worthy of his attention.

As he passed the cabin adjacent to number fourteen, he heard unmistakable sounds coming through the closed windows. He smiled grimly to himself, knowing what it meant. Hasad had driven a hard bargain for the girl, and must have forced Cari to screw him first. The fact that Cari was actually going through with it meant that she too thought the American worth so heavy a price.

His interest now was even keener than before. He slipped into the observation storeroom, put his eye to the peephole, and nearly gasped out loud.

The American girl was truly a goddess!

He could tell that she had been drugged by the manner in which the blonde young innocent was desperately trying to satisfy herself as her fingers massaged her clitoris and slipped wetly in and out of her openly exposed cunt. Onan licked his lips at the sight of the long tapering legs, rounded white ass-cheeks, firm, pink-tipped breasts. He strained to make out the girl's face, but though the outline was promising, he couldn't quite see her individual features.

As he watched her body thrashing and writhing on the bed as though being fucked by an invisible man mounted over her bucking torso, he felt his cock jerk inside his shorts, the beginning of an erection already making his balls ache with lust.

He had made up his mind. If this girl's face matched the rest of her, he had to have her. And there was only one way to find out, risky as it was. He closed the peephole and went out of the observation room toward the door into the cabin.

Luckily, he thought, the key is here. Making as little noise as possible, he let himself in and edged his way carefully toward the bed.

Inside the room, the moaning noises the drugged young bride was making filled the air as she strained in her semi-conscious state to satisfy the craving heat of her wide-splayed loins. The sense of desperate longing in her voice made Onan's cock lurch heavily and thicken into a protruding bulge in his pants. He leaned over her thrashing form on the bed and looked down into her lust-contorted face.

Her eyes were tightly closed and her lips

compressed into a narrow smile of sensuality, but he could tell that she was the most beautiful young woman he had seen for a long time. By now his expanded cock was throbbing painfully from the sight of the lewdly masturbating young blonde, and he unzipped his fly and squeezed its heavy pulsating length in his hand. This somewhat eased the aching, but he knew that there was only one way to rid himself of the bloated feeling building in his swollen testicles. He pushed the waistband of his shorts down inside his pants allowing his blood-distended cock to stand out in front of him, protruding over the naked girl on the bed below.

Slowly and experimentally he stroked the loose outer skin up and down over the blunt tip-end, considering whether or not to cum that way on the heaving white breasts of the young woman so wantonly lost in the throes of passion beneath him.

Then he kneeled on the bed beside her, running his free hand reverently over her firmly ripe breasts, her flat white belly, her tautly moulded thighs – all over her unprotected nakedness – his gentle caresses bringing forth small animal mewls of pleasure from between the lovely blonde's clenched teeth. His passion increased as he watched the contrast of his fully clothed arm moving over the creamy-white smoothness of her body, bringing a gasping, quivering response from deep inside her chest.

He watched her open her eyes, their blueness shadowed by the dim smoky veil of passion that was coursing through her veins as she looked up at him, a dazed expression on her beautiful face.

Somewhere back in the distant ages since she had felt her husband John's hotly moist mouth slaving at her undulating loins, Lucy had drifted into a strange unknown world of sudden deep, soft pleasure of the flesh. Now all conscious thought of why she was here or who it was bending over her flaming body was lost to her mind, for it no longer mattered. She felt the tiny licking fires running all through her, between the softness of her inner thighs, out the tips of her now throbbing pebble-hard nipples, and down again to the burning core of her vagina where it roared in white hot heat like the interior of a furnace.

"Oh God," the drugged young wife whispered softly, unaware of what she was saying. "Oh God, you make me so hot, honey."

Unable to resist, Onan pressed his wetly searching lips down over hers and immediately felt the hard probing pressure of her tongue spearing up into his mouth in her unconscious quest for a warmer, wetter contact with the devil-like thing that had been so deliciously torturing her body before.

The dark-eyed slave dealer's hands continued exploring her body. When he pressed them up between her thighs, she did not jerk away as he had thought she might, but voluntarily opened them wider, admitting him to the very core of her being where he could feel the warm moist centre of her loins flexing in passionate answer to his probing touch.

"Aaaah, aaah," she was sighing continuously up into his lips, her tongue pushed up against his, circling deep around inside as though searching desperately for another entrance to a deeper part of his mouth.

Lucy was ecstatic that her bridegroom had relented from his anger with her and had returned to placate the craving she felt gnawing at her belly like an insatiable voracious animal. She couldn't wait much longer for him to come to her and quench her unending thirst.

Her passion spurred Onan on even though, knowing Cari and Hasad to be somewhere nearby, he normally would have left the premises immediately if only to preserve his cherished image of self control. Amazed disbelief buzzed in his brain that this could be happening. He had seen many young innocent women, even virgins, brought to this peak of uncontrollable sensuality, but he had never stepped into a room and found such an incredible beauty writhing and twisting in lewd abandon as though she were just waiting for someone to crawl between her legs and fuck up into her. This was too much for him to ignore.

Unable to stand it another minute, he moved over her, his fully clothed body covering her warm soft form like a protective blanket. He could feel her shuddering helplessly out of control under him and, grasping her unresisting thighs in his fingers, he drew them up until the whole of her wetly throbbing vaginal slit was presented up to him in welcome sacrifice. He held them there a long tortured moment, and then finally reached down to implant himself in the widespread opening.

The lust-possessed young bride lay groaning in tiny unintelligible gasps that hissed out from between her gnashing teeth like quick puffs of steam from a labouring locomotive. Her body ached and she was vaguely aware of a heavy dark form hovering over her as her legs and loins waited, quivering and wet, hungering to be filled.

Onan let his breath out slowly as he felt her hand burrowing down without hesitation between their bodies until her warm fingers closed greedily around his rigid cock-flesh. He felt her thumb and forefinger tighten around the head of his pulsating penis, slowly but firmly shoving his own hand away. He gasped and his lips bared back off his teeth as her hand gently pressed the moist sensitive tip of his cock into the fleshy wet furrow of her hotly palpitating pussy. He could feel the soft blonde pubic hair grazing gently against his tingling cock-shaft as it hung poised for entry between the excitedly throbbing cuntal lips.

The broad rubbery tip resting at the narrow opening to her cunt increased Lucy's overwhelming hunger until she thought she would burst. It had seemed so long, so long that she had waited for this, and now she was about to have it.

He paused another long moment enjoying the warm moistness of her pussy quivering in desperate anticipation around his eagerly pulsing penis-tip and then – with one long smooth stroke – he drove it into her all the way up to her womb. There was not the slightest resistance now after the long desire-crazed build-up she had gone through, but he could feel that she was narrow and almost virginal, the smooth wet walls of her cunt clamping around his cock so snugly and firmly it almost drove him out of his mind.

"Oooohhhhh," the ecstatic blonde moaned in a fluttering voice as she felt her entire insides open to receive him deep up in her belly. She began groaning and murmuring incoherently as he fucked viciously into her, gritting his teeth at the excruciating hotness of her voraciously clasping vaginal passage. Her body followed his, matching his strokes with wild abandoned jerks beneath him.

"Ooooh, God. Ooooh God, so good, so goooood!" she groaned again and again, her arms wrapping tightly around his neck, pulling his body closer to hers until the buttons of his vest dug into the swelling whiteness of her firmly resilient breasts.

He plunged his long thick cock up and up, deeper and deeper into the warm soft cavern of her cunt, feeling the whole of her belly flowering open before his relentlessly pummelling onslaught as though she had waited for him all her life. Her entire body bucked and twisted as she moaned again and again, her face contorted in ecstatic passion below him. Her mouth moved ceaselessly and her nostrils flared in the untamed animal desire that had taken hold of her body. Her forehead was covered with a light mist of perspiration that glistened and soaked her dishevelled long blonde hair. There was nothing that could stop her wild race for fulfilment now and Onan fucked like a madman to bring to the peak of soaring climax.

"Ohhh God, yes, yesss," she cried up into the room as he slithered his wriggling sinewy hands up under the wildly pumping cheeks of her buttocks, cupping them tightly to raise them up off the bed for greater access to her open and pleading loins.

The lust-inspired Turk squirmed down into her from that position with all the strength of his hips and thighs and could feel the smooth raw flesh of her cunt clasping and unclasping like a heartbeat all around his hot, near-bursting cock. He fucked into her with a power summoned all the way from the tips of his toes, ramming the last thundering inch of his thick rod up into her churning belly, bringing new ecstatic moans from her lips that resounded through the room like the cry of a wounded doe. Her nostrils flared again and her eyes, open wide now, gazed glassy-eyed and unseeing up into the semi-darkness as they burned lustfully with a wild, unsatisfied desire.

He pulled his head back so that he could watch her contorted features. He didn't want to miss the facial expressions of sudden and humiliating total surrender, the abandoning of this young woman's whole being to the sensations of lewd sensuality. Her lips bared back over her teeth, with greater and more desperate, guttural sounds emanating from deep within her throat.

Her arms, which had been wrapped tightly around his neck, slithered down now over his back, and her hands dug demoniacally into his muscular thrusting buttocks as they pounded down into her open unprotected loins, tearing at the material of his trousers with her nails as though she wanted to rip them off and fling them away. The zipper of his pants caught and pinched at the tender skin of her upturned pussy and she jerked wildly under him from the pain, then caught up the rhythm again and surged once more into the race for completion. Wet sluicing noises resounded through the room with each cruel driving lunge he made into her, blending with the soul-shattering groans coming from her throat.

Lucy felt his hands run from the smooth hollowing cheeks of her buttocks down to her inner thighs and between them to the moisture soaked pubic hair surrounding her cunt. He rubbed his fingers into the soft fleshy flanges which clung and throbbed around his plunging cock-shaft like a rubbery wet mouth, and the maddening sensations it caused to quiver through her bucking thrashing body drove her closer to the brink of madness. Her body was slippery from the sweat of her invitingly untamed gyrations, and her head flailed uncontrollably back and forth on the mattress, her long blonde hair lashing and whipping like a cat-o-nine-tails. She opened her mouth gasping for breath in her wanton passionate abandon. She had become something crazed and inhuman, twisting and jerking, spreading her legs wide apart and wriggling them against his down-pressing shoulders, urging him on.

"Oh, oh, deeper, harder, fuck me, *fuck me*," she begged in words she had never used before, grunting as if she were slaving with her body to punish his endlessly pumping loins. She was on the very verge of her orgasm, and she swung her feet straight up at the ceiling behind his head, her toes curling into tight little balls as she strained toward the final release that had been building in her for so long. Her ass-cheeks were undulating and waving uncontrollably from side to side under him as she spiralled her cunt wantonly up and down on his moistly plunging cock.

"I'm cumming!" the drugged young blonde suddenly squealed as if in surprise. "I'M CUMMMMMING!" With a high-pitched gasp of intense passion she locked her ankles in a death grip high up behind his labouring back. Her body arched up at his and her arms snaked tightly around his neck, her mouth pressing suffocatingly over his lips to suck and lick at them in devouring hunger.

Suddenly she was locking her whole body against his like a leech, not moving but quivering and quaking against him in a pulsating hula rhythm as she spewed her cum-fluids out around his still hard-driving cock and down the wide-split crevice of her buttocks.

"Fuck, fuck, fuck," Onan repeated softly over and over again, like a mantra, into her hot moist mouth, picking up the lewd words he had heard spouting from her lips and using them to spur her on. He felt the warm juices seeping around his teasing fingers on her cuntal lips, and he suddenly darted out with his moisture-soaked middle finger, thrusting it without warning into the tiny puckered ring of her anus. She moaned up at him, bucking painfully from the sudden unexpected entry, at the same moment as he felt a slow aching pressure soaring deep within his heavy balls. Gripping the cheeks of her insanely rotating buttocks hard with his outspread fingers, he squeezed the firmly moulded flesh tight around his finger extended deep up into her anus, feeling her cringe and vainly attempt to jerk away from the invading pressure in her rectum.

Then, suddenly, as though great tidal waves of heat were surging through his veins, he felt himself cumming and increased the tempo of his fucking strokes to a machine-gun pace, grinding with each thrust as far up into her open and yielding cunt as he could go, bringing groans of post-climactic passion from her lust-wracked chest. He moaned like a death-rattle into the wet cavern of her mouth, felt the bursting in his loins, heard her whimpering cry and the tightening of her arms around his neck, and then, with a deep mind-shaking gasp that went on and on into her mouth, he exploded hotly inside her, shooting his thickly spewing sperm in endless streams up into her receptive cunt. He felt it spurting again and again, each pulsing gush wracking his whole body as if it would tear it apart, the hot cum ricocheting inside her clasping vagina filling her belly and mingling lewdly with her own flooding juices around his ejaculating cock.

They lay still then, their tightly embracing bodies locked together, throbbing and quivering like animals huddled tightly in a ball after a furious mating in the untamed wilds. After what seemed an eternity, the drug-controlled blonde suddenly collapsed away from him down onto the mattress, her arms and legs flung limply out to the sides.

Her hungering need had at last been fed, and she fell into the deep exhausted sleep of contentment that follows the satiation of all bodily needs.

Onan felt his depleted penis slip wetly out of her relaxed moist cunt, and reluctantly lifted himself away from her spread-eagled young body. He would have liked to have brought her to life again and taken a great deal more pleasure from her obviously newly awakened passion, but he knew that he had to make his move fast. He had already stayed longer than he had intended to, but luckily Hasad and Cari had not finished what they were doing. Moving with quick, cat-like speed, he left Lucy alone and went to his car. He told his driver, Mahmet, to go back, wrap her up in the bedspread and carry her to the limousine.

There will be plenty of time to enjoy her at my *yali* in Istanbul, he smiled to himself. And there I can dally with her at my leisure without any danger of being surprised in the act. He looked forward to teaching her all the things he would want her to do, and he congratulated himself for getting such a prize without paying one kurus for her.

* * *

John ran from the hotel into the noisy streets of Istanbul, cursing himself under his breath for his senseless inconsiderate treatment of his young bride earlier in the evening. He had been needlessly brutal and unfeeling toward her, and now she had run away from him in this foreign, sinister-looking place. If he had it to do over again, it would have been different, he told himself, but now it was too late.

He had returned to the hotel to find her gone, and after questioning the night porter and bribing him well, John had finally learned that Lucy had taken a taxi. The porter said he had overheard her telling the driver to take her to a motel on the Dardanelles out in the country, and John was determined to follow her there and plead with her to give him another chance.

He was sure that she loved him, and if only he could find her she would let him try to make up for his mistake.

It seemed to take forever for the ferry to get them to the other side of the water that divided the huge city into two parts, but they were finally speeding through the night along the rough roads toward his abused runaway bride. They passed very few cars as they wound into the country, but once they nearly crashed head-on into a long black limousine at a dangerously narrow curve. He caught a brief glimpse of the well-dressed man sitting in the back seat calmly smoking and stroking the head of a woman sleeping against his shoulder. John shouted incoherently at him, shaking his fist in anger at the careless way the man let his chauffeur drive. But the face only smiled insolently at him as the limousine sped on toward Istanbul.

John's driver had indicated that it would take at least an hour to get to the motel once they were free of the city traffic, and he settled back nervously into the seat, smoking one cigarette after another.

* * *

At first Hasad found it hard to believe that the American girl was not in the cabin where he had left her. Then he spotted the half-smoked oval cigarette Onan had inadvertently left in the gravel outside the door, and he knew what had happened. He decided that Onan must have seen Cari's car waiting outside the office, and angry that the double-dealing Hasad was trying to get away with something, that he had kidnapped the girl while Hasad was screwing Cari. He wondered if Onan would ever pay for the young beauty.

When he told Madame Cari that the girl was missing, she became furious at her blackmailing lover, suspecting that he had planned the whole thing. After having his way with her, she thought, he had gone ahead and permitted the wealthier Onan to take the young woman away to Istanbul.

It wasn't true, but there was no way to prove it to Madame Cari, and she left saying that he would have to find someone else to deal with in the future. She then slapped his face, promising to get even with him in some way, and left the motel in a fury.

Hasad was alone in the office fuming over his misfortune when the American's husband, John Dean, had come in demanding to be told where his wife was. Hasad was taken aback, not knowing for certain until now whether the young woman's husband was, in fact, in the country. After trying to stall by saying that he had never heard of any such woman, he realized that the man was the type who might contact the police. It was then that Hasad decided that he had probably already lost any chance of future dealings with Onan Bey, so he abruptly decided to tell this John Dean where he could find his wife.

Hinting that she had come with another man and spent some time with him in one of the cabins, he told John that his wife had found the motel too isolated and lonely and had decided to go back to Istanbul, where she intended to check in at the Palas Hotel. Hasad knew that the old wing of the hotel was now one of Onan's several properties in Istanbul, and he had no doubt that that was where the girl was to be found. It gave him at least a small sense of satisfaction to know that he would be causing some difficulty for Onan who had robbed him of his deserved money, not to mention the pleasure of fucking the young wife before she was taken away.

It was only after John had left to return to Istanbul, promising trouble if he didn't find his wife where Hasad had told him she would be, that the vengeful Turk remembered that he had the girl's passport still hidden in his blazer pocket. He ripped its pages into tiny shreds and burned it in a wastebasket, destroying all traces by crumbling the charred ashes between his fingers. Now let them prove she was ever in one of these rooms, he snorted to himself.

To be doubly sure, he went to number fourteen and cleared away all indications that anyone had been there that night. He looked with regret at the bed where he had been on the point of screwing the drugged young bride only a short time before, and then he locked the cabin and went back to the office. He still had time for a quick nap before the morning crew came on duty, and he needed it after the wild scene with Madame Cari.

At least the evening hadn't been a total failure, he grinned obscenely to himself. He kicked off his shoes and stretched out on the couch inside the office, and within a few minutes he was sound asleep and snoring.

Chapter 5

The Palas Hotel was located in one of the older sections of Istanbul with narrow, cobbled streets. John Dean watched the seedy neighbourhood his taxi was entering with some anxiety, noticing the dark, surly men skulking in the doorways opening right onto the streets and the painted women who were obviously prostitutes. Finally the car stopped outside the entrance to a surprisingly modern looking hotel, and the driver asked to be paid.

After a brief argument that ended with John paying almost double what he thought he should have, he got out and ascended the broad steps. He brushed away the five dwarfs in red circus-like costumes who scrambled down to meet him. He had not brought any baggage, leaving his belongings in the hotel he had checked into so many hours ago with his bride Lucy.

At the desk he demanded to see the manager and someone who could speak English. He was exasperated with all the difficulties he had been encountering, and he was so tired that his patience with these people was wearing thin. The man at the desk pretended not to understand a word he was saying, and John began to shout and bang his fist on the reception desk. He was almost sure Lucy was in the building after what that sneaky looking little night manager at the motel had told him, and he would find her if he had to search every room himself.

John had not noticed he was being watched. The pretty dark-haired girl standing in a corner of the room glanced at a photograph she held in her hand until she was finally satisfied that the man shouting at the hotel clerk was the same as the one in the wedding photograph.

Onan had taken it from the handbag where Lucy had put it after her mother had handed her the Polaroid snapshot outside the church. It showed her and John in their wedding costumes, smiling at each other on the steps under a shower of confetti and rice. The picture had been definite proof that the girl was married, and Onan was sure that her husband would be desperately searching for her. The experienced slave-dealer had no illusions about Hasad's ability to figure out what had happened to this blonde American, and he had posted the girl Akasma in the hotel lobby to watch for John.

The desk clerk was nervously wringing his hands and John was threatening to call the police when Akasma stepped across the room and touched him lightly on the elbow.

"Please, would I be able to help you, sir?" she inquired in a sweet voice. "He really doesn't understand a word you're saying. I'm afraid that few of my countrymen speak your language."

John turned around in surprise at the sound of the soothing feminine voice. He looked down into the softest, warmest pair of brown eyes he had ever seen. His anger caught in his throat and he changed his tone of voice.

"God, yes. Thank you very much, Miss ... Miss ..."

"You can call me Akasma," she answered him demurely, gazing shyly into his eyes.

"Thank you, Miss Akasma," John went on, stumbling a little over the strange-sounding name. "I was telling this clerk that my wife is in this hotel, and I want to know which room she's in." He knew it sounded crazy to say that, but there was nothing he could do about it. It was the truth.

"Just a minute," she told him, reassuringly patting his arm. "I'll get the information you want. What is your wife's name?"

"Lucy... Lucy Dean," he stammered. "And I'm John Dean," he added.

"I'm pleased to meet you, John," she said warmly, smiling up into his eyes. Then she turned to the clerk and spoke Turkish rapidly but with a softness John had not heard in the language before. He watched her full sensual lips forming the words as though she were caressing each one with her mouth. While she was busy asking questions and listening to the answers the man was giving her, John took the opportunity to run his eyes appreciatively over the young woman's body. She wore a clinging silk mini-dress that left little of her ample physical attractions to the imagination.

Her high full breasts thrust tautly out against the material of her dress, lightly rising and falling as she spoke. A tiny, almost invisible belt tied the dress in at her narrow, wasp-like waist, accenting the outward flare of her hips and well-rounded buttocks. He followed the long tapering line of her legs below the short skirt to the enticing swell of her calves and her slim ankles. She's quite a dish, he thought to himself, forgetting for a split second his concern about Lucy.

He glanced at the profile of her face again and was nonplussed to see her turn to smile coquettishly at him. She had seen him running his eyes over her body, but it didn't seem to embarrass or upset her. In fact, he thought, she seemed to like it. He smiled back a little more boldly than before, thinking to himself how lucky it was that she was here when he needed help, and that she was also lovely to look at.

"He says that your wife is not a guest of this hotel," Akasma told him softly.

"I don't believe it!" John nearly shouted again. "She was brought here from outside the city not more than a couple of hours ago. Tell him that I know she's here!"

Akasma placed her hand soothingly on his arm and squeezed lightly with her fingers. "I understand how you feel, John. But he swears by Mahomet that she's not here – at least not under that name. Perhaps she isn't alone?" she suggested, changing her tone slightly.

"Impossible! We were just married yesterday!" he blurted out. Then suddenly he remembered the sly hint the man at the motel had made. It was just barely possible that Lucy had actually been with another man – perhaps she was making a desperate attempt to get back at him for doing and saying those things earlier in the hotel – especially his declaration that he was going out to find a good lay. Suddenly he didn't know what to think any more. "I just don't understand women," he admitted.

Akasma noticed his sudden uncertainty and leaned over closer to him, her breasts, seemingly inadvertently, brushing against his upper arm. "Let's go away from the desk and talk alone for a minute," she whispered in his ear.

They moved to the other side of the room and she planted herself firmly in front of him, looking sincerely up into his eyes. "I can see that you and your new wife must be having problems," she said. "I'm sorry. I didn't want to say anything over there, but it seems to me that the clerk was hinting that there was a woman who might have been your wife who came in a little over an hour ago – with a Turkish man."

She watched the glowering expression

that seemed to close down over his face like a dark cloud. These Americans are so easily swayed in their emotions, she thought. She was glad that he was good looking; it made her job so much more interesting.

"If it's true," John began in threatening tones, "just wait till I get my hands on her again!"

"You're tired and upset. Don't come to a hasty decision until you know for sure what's happened," Akasma tried to calm him. "I have a room here. Why don't you come up and rest, and perhaps I can make some discreet inquiries among the hotel staff."

John felt his shoulders drooping from fatigue and worry, and, not knowing what else to do, he gave in to the light pull she gave his arm and followed her into the hallway. At least I won't be alone for a little while, he consoled himself. He couldn't help but think as he watched her seductively swaying hips in front of him that she seemed more interested in him than just to help find his wife. He hoped so – he would love to get his hands on her appealing young body.

She led him through complicated winding corridors until finally they entered an older part of the hotel. Eventually she stopped outside a door and fumbled for the key hanging from a thin gold chain around her neck. She unlocked the heavy oak door, and

John followed her inside. He was amazed at the opulent luxury of what seemed to be a large private apartment.

"You like it?" Akasma asked him, watching his face with interested amusement. It was obvious he had never seen such lavishness except in movies. The floor was covered with thick Persian rugs, and all the furniture was upholstered with highly ornate Moroccan leather. The low tables were covered with objects of chased brass and gold left over from wealthier days when sultans ruled the country.

"This is what I call living in style," John managed to blurt out. "Are all the rooms here this luxurious?"

Akasma laughed deep in her throat, a husky sensual sound that made his loins ache with desire. Here was a woman he could really get to like if he were given half a chance.

"No, this is my own special apartment," she answered with pride. "It used to be for a rich man's favourite mistress, but I have lived here nearly five years now."

John wondered how a young Turkish girl could afford such an expensive-looking place, but he didn't ask any questions. It was possible that she, too, was able to stay there only because she was the kept woman of a wealthy man, and she might be sensitive about it.

"Make yourself at home," she told him. "I'll be back in a minute." Then she disappeared into another part of the apartment and John amused himself by looking over the room. A beautiful, ornate hookah was near the low-lying couch, and he wondered briefly what, if anything, Akasma smoked in it. He finally decided it was probably just a decoration. He ran his fingertips over the thick animal skins scattered around on the furniture, enjoying the sensuous quality of the furs.

Before long he began to wonder what was taking her so long and sat down on one of the divans. He suddenly realized just how tired he was and leaned back to rest his eyes. For a moment he lay there letting his mind go blank, but then again he found himself wondering about his lovely young bride. He hoped that she was all right and wished vaguely that he had been able to catch up with her before this. Then he was suddenly annoyed with her again for disappearing like that, and after deciding that he had done as much as he could for now, he relaxed and dropped off into a light sleep.

* * *

In another part of the huge hotel, Lucy was slowly recovering from the effects of the drugs Hasad had put into her drink earlier. Onan had turned her unconscious form over to an old woman-servant who was plying the dazed young wife with thick black coffee while she lay soaking in a sunken marble bath filled with hot water and attar of roses. She felt completely drained and utterly exhausted. However, the sleepy blonde was slowly returning from the hazy dream world, where she had languished for what seemed an eternity, to the world of reality. The hot coffee and the warm water were helping to revive her, and now that she was regaining consciousness she was becoming more and more aware of her surroundings and beginning to wonder where she was.

The old woman held up a hammered copper coffee pot, but the confused young wife made a sign to show that she had had enough and the servant bowed and silently left the room.

Lucy took the opportunity to crane her neck and look around. She was trying to understand how she had arrived here. The last thing she remembered was getting into bed back at the motel, and after that it seemed she had been dreaming and dreaming for ages. She couldn't help but recall the previous events that had led to her running away from John and the hotel bridal suite, and she wondered about her husband – she felt a little sorry for him. Probably he was sick with worry. Well, she thought, let him worry. Tomorrow, she would return to their hotel and maybe by then John would have learned a lesson. She had decided the whole thing was silly. If only he would try as hard as she would, surely they could work it all out.

She peered curiously from the tub out through the open arched doorway. The building looked like a palace from where she was sitting, and for a split second she considered the possibility that she was still dreaming. Otherwise, how could she explain her presence here? But she could feel the warm wetness of the water on her skin and still taste the cloying sweetness of the coffee in her mouth.

No, she definitely was not dreaming. Then what had happened? Could it be that John had found her at the motel, and somehow, without waking her, had moved her to this luxurious palatial apartment? She was suddenly sure that was it, and she was about to call his name, when she heard softly padded footsteps in the other room. That's probably him now, she thought, coming to surprise me with his cleverness at finding me after I ran away without leaving even a note. She smiled to herself and expectantly watched the doorway.

"Hello, my dear. Are you feeling better?" Onan asked.

Lucy gasped and shielded her exposed breasts with her hands, sinking down into the bubble bath as far as she could.

Standing in the doorway was a man in a flowing djellaba – a totally unfamiliar man. He was smiling at her in a relaxed and familiar manner, and for a moment she wondered if she *should* know him. Had she had too much to drink and somehow allowed herself to go off with a perfect stranger? She stared in fascinated horror at his unusually handsome face, taking in his deep-set dark eyes, full sensual-looking lips trimmed with a thin moustache, and noticing the large gold signet ring that he wore. No, she was certain, she had never to her knowledge seen him before this moment.

"W-who are you," she stammered, her voice trembling as he gazed knowingly down at her as though she had belonged to him forever.

"Onan Bey," he answered, bowing slightly. "Your new master."

The abducted bride's mouth gaped open at his words. It had to be some kind of weird joke – she had never heard of such a thing in her life.

"You seem a little taken aback, my dear," Onan said in a tone of superiority. "You mean to tell me that your husband John said nothing at all to you about our little arrangement?

You see, you are now my little slave: you must do everything that I tell you to do." He watched her face with an inner chuckle as her mouth fell open in undisguised dismay, eyes widening in disbelief at his meaning.

"B-but he couldn't do such a thing!" she sputtered. "It-it's preposterous! And besides, John loves me," she added as an after thought. Immediately after her words had left her lips, she wondered if it were not possible after all. This part of the world might have totally different laws from other countries for all she knew, or this could be happening without the knowledge of the authorities. And there was no doubt that John had been highly displeased with her when he had left the hotel.

"Are you sure?" Onan goaded her teasingly. "He certainly seemed happy enough to accept my offer, and the price was not too steep, either." He was enjoying playing with the confused young woman like a cat with a mouse. He knew how proud and independent these American women were, and nothing pleased him more than to make them feel like helpless slaves who could be bought and sold on a mere whim. They usually changed their attitudes once they understood that they would be treated with the respect that normally fell only to royal concubines. All they had to do was admit that they were at the mercy of his will, and life could be extremely good for them – at least as long as he wasn't bored.

Lucy stared at his grinning face for a long time, shaken by his calm, secure manner. Finally she had had enough.

"Please leave the room," she said curtly. "I'm getting dressed and I'm leaving." She jerked her head around and stared blankly at the tiles on the wall, waiting for him to obey her command. But he only laughed softly and remained standing in the doorway.

"I'm afraid you'll have to change your attitude, my dear," he said after a long pause. "You don't seem to understand. You are now completely at my mercy, and you will have to do as you're told. Never again will you be able to order anyone around – except the handmaidens who will take good care of you for me. Even they will listen to me first if there is any doubt about what to do." With that he snapped his fingers once, and she heard the shuffle of feet behind him.

Two young girls dressed in loosely fitted white robes appeared with thick towels draped over their arms and a bundle of clothing clearly intended for Lucy. Onan rasped a few brief words to them in Turkish, and they crossed to the tub where Lucy was still trying to hide her body from his view.

"These women will assist in preparing you to come to me," he told her coldly. "Do not give them any difficulty. My man Mahmet is waiting outside, and he's very anxious to begin teaching you proper respect. He believes that women are like animals – both need a little training before they behave the way they should." Then he was gone.

The first woman unfolded a huge towel and held it out for the completely astounded young blonde. Lucy hesitated a moment and then decided that she would have to dry off and dress before she could do anything about her situation. She rose up from the water, slightly embarrassed at exposing herself to these total strangers – even if they were her own sex.

They seemed not to pay the slightest attention to her nakedness, rubbing her tender skin dry with great care and skilfully plying her body with sweet-smelling oil as she stood between them like a lovely princess of old. In spite of her fear and confusion, their hands felt good on her soft skin and soothed her nerves. When they were finished they stood on either side of her and began to drape her almost reverently with a beautiful dark green silk gown embroidered with tiny gold tulips. Lucy happened to glance in the full-length mirror at the far end of the room and saw herself as her shimmering nakedness slowly disappeared beneath its caressing folds. Her tall lithe body was softly glistening

in the muted light, her breasts gleaming like twin white moons, and her full flaring hips and tapering thighs had never looked so appealing. Even the golden triangle of pubic hair up between her legs shone from the oil, matching the loosely falling locks making a blonde parenthesis around her face. She watched in wonder as the two women cinched in the waist of the robe to cover her body, finishing their work by slipping her feet into thin gold sandals and then bedecking her with large, circular gold earrings and numerous matching bracelets.

The young wife couldn't help admiring herself in the mirror, and she had to grin at the childish way the girls stood back and waited for her to compliment their skill. She felt like Cleopatra, quickly flashing a smile at them before she strode regally out of the room.

The man who called himself Onan was nowhere to be seen, and Lucy curiously explored the first room she entered. It seemed to be a kind of dressing-sitting room, with ornamental mirrors and vanity tables, many overstuffed chairs and lounges, and rows and rows of bottled perfumes and powders. She opened a wide closet and found it filled with all styles of dresses – a wardrobe such as she had never seen except in movies and magazines. They all seemed to be about her size, and she was secretly envious of whoever owned them all.

Lucy walked into the next room and found in the middle against one wall the largest bed she had ever seen. It was covered by a red velvet canopy hanging from four carved posts at the corners. Stepping closer, she realized that the bed was surrounded by mirrors made to look like room dividers, with a circular mirror mounted in the canopy over the bed. She knew that there could be only one purpose for them. Anyone on the bed would be able to see himself from whatever position he was in.

This man Onan liked to see himself making love to women. She had heard of such things in naughty whispered conversations, but never had she dreamed that they actually existed. She must have been brought to a virtual den of iniquity – and she had to find a way out before it was too late.

She rushed over to the window and looked into a huge high-walled garden. It would be easy enough to slip out the window and try to find a way over the wall, she was thinking, but then she saw the dogs and her heart seemed to leap up into her throat.

They were almost as tall as her waist, with long pointed snouts and sharp teeth. Obviously, this avenue of escape had been tried before, and Onan wasn't taking any chances. She wouldn't last five minutes outside with those hungry-looking beasts roaming restlessly under the window.

Then she tried what she hoped would be a door leading to the hallway, only to find that it was locked. She was trapped in these rooms, and she began to wonder if she would ever get out. Suddenly the door she had just tried opened and Onan came in. Lucy caught a glimpse of a giant man standing guard outside the door and fought down her impulse to make a run for it. That was probably the Mahmet he had mentioned before, and he looked like he could snap her in two if he wanted. She drew in a deep breath and straightened herself into a stern pose of indignation.

"All right, Mr. Onan," she began. "Thank you very much for your kindness, but now I demand that I be allowed to have my own clothes." Lucy quickly slipped off the gold bracelets and the earrings. Then placing them on the chair beside her she said, "I must leave now. My husband is expecting me." She hoped that he wouldn't see how frightened and hopeless she felt inside.

He totally ignored her words.

"I see you've inspected your suite," he said in a mild voice. "I hope it meets with your approval. You may be spending a lot of time here."

She opened her mouth to speak, but was

unable to form words with her lips. He was taking it for granted that she would stay here until he decided she could leave.

"Here is something which might interest you," he said in a shrewd, suggestive tone. He opened a pair of louvered doors to reveal a large television screen. Flipping a switch, he motioned for her to sit in a chair facing the screen, and then he adjusted the dials until the picture became clear.

She hadn't intended to sit down. She wanted to show him that he couldn't tell her what to do, but when she saw what was on the screen, she practically fell back into the chair from shock.

There was John – with another woman!

"This is coming from another suite in the building," Onan told her in a flat, cold voice. "You can decide for yourself whether or not your husband is concerned about you."

With a smile of cruel satisfaction, Onan lit one of his cigarettes and stood back to enjoy what was coming. He had just talked with Akasma, and knew that the American was ready to do anything she might ask of him. They had smoked some strong hashish in the hookah, and the young husband was already beginning to make advances when Onan had called her away to briefly talk with her. Now they were together again, and Akasma had been instructed to go ahead.

Lucy barely noticed when Onan reached to one side and flicked off the lights in the room, and then pressed a button which caused the heavy window drapes to glide closed, leaving the television as the only source of illumination.

On the screen Lucy could see a large double bed almost like the one she had just been looking at. In the center of the bed a long-legged, dark-haired girl was lying spread-eagled, completely naked before the hidden camera. John, also stripped bare, was kneeling on the bed between her widespread legs running his lips and tongue excitedly up and down her body. He sucked at her ripe, rounded breasts, teasing the nipples up into trembling hardness, and then he traced a path with his tongue over the flat plane of the girl's soft belly all the way down between her quivering inner thighs.

Lucy gasped out loud at the obscene pleasure John was taking over the lewdly prostrate form on the bed. If that's what he wants to do to me, he's got a surprise coming, she was thinking. She had never seen such wanton depravity in her whole life.

Then on the screen Akasma's hands tangled in John's hair, pulling his greedy lips tightly to her loins. Lucy leaned forward gripping the arms of the chair until her hands ached. She could not move. She tried to close her eyes and blot out the shocking picture but she couldn't even do that. She had to look – the salacious spectacle was hypnotizing her even as her mind rebelled – to watch her own husband toying with the naked squirming body of another woman was debasing and lewd. Still, she could not turn away.

Lucy could see that her husband had worked the dark-haired Turkish girl up to a fever pitch and her mouth hung wide open in ecstatic rapture, her glassy, lust-crazed eyes staring up into nothingness. He placed himself in position over her rapidly rising and falling belly, his face hanging within inches of the black-furred pubic mound below. Her fingers still clutched in his hair and struggled desperately to pull his face down to her squirming cunt below, but he was obviously holding back until he was ready. His hands pressed down on the smooth flat plane of her lower stomach as his thumbs squeezed into the fleshy outer flanges of her vaginal lips. Lucy sucked in her breath with a gasp as her husband's thumbs massaged the other woman's soft moist pussy-flesh for a moment, and then pulled outward slowly, parting the dark wet curls to expose the moist inner folds of her cuntal crevice. The tiny throbbing bud of her clitoris was clearly visible just above the stretched elastic vaginal opening.

Then John finally gave in to the straining

hands in his hair and his head dropped, his probing wet tongue snaking out to teasingly flick at the quivering little clitoral nub. Akasma's body jerked at the sudden electric contact and her legs clamped tightly together around his head, the firm softness of her inner thighs imprisoning his ears in a vice-like grip. Her hips began a slow up and down movement in rhythm to the ministrations of his lashing tongue as soft mewls of animal pleasure came from between her clenched teeth. She was lost in a mindless uncontrolled lust as her upper torso writhed like a belly dancer's, her wildly gyrating buttocks grinding spasmodically down into the mattress.

Lucy watched the lewdly depraved scene as though it came from a distant dream she had once had, strange chords of remembrance passion through her head. It was almost as though it had all happened to her somewhere in the past, but she was unable to believe it possible that she could ever react in such an abandoned way. Her mouth gaped in disbelief as she watched Akasma twisting wantonly under the degrading sucking at her loins. The confused young blonde was no longer certain whether the two writhing bodies she was watching were real or a figment of her imagination. It felt as though they were dancing in her head. Time and distance had suddenly become indefinable. Nothing existed in the world but the scene before her, and she seemed somehow mystically connected to the two figures on the screen.

It suddenly dawned on her that when she had first been brought to this apartment, not even aware that she had left the motel, it was entirely possible that Onan had performed similar lewd acts on her own helpless body. It would explain a lot of things to her confused mind. He seemed to consider her his property even though, to her knowledge, up to this point, he had never even touched her. And she had felt a little strange and sore down in her genitals when she had first come out of her stupor of sleep.

The embarrassed young bride suddenly blushed in her new understanding, and she glanced sideways at Onan standing beside her to see if he had noticed. But he was staring straight ahead at the screen as if lost in the lascivious act taking place there, a thin smile curling his sensual lips.

She glanced back at the screen and saw that John had stopped what he was doing at the squirming woman's open loins, and now he was crawling up her writhing body, his lips slowly working their way up, pausing now and then at soft sensitive spots for a quick flick of his tongue. It was then that Lucy saw John's distended erect penis protruding down between his legs, the heavy testicles beneath filled to bursting with sperm. The massively rigid cock paused momentarily over Akasma's undulating vagina, and Lucy thought that her husband was going to ram it up into the girl right then. But instead he slowly moved all the way up her slippery perspiring body and straddled her firmly upthrust breasts.

Lucy could see the proud peaks of her breasts squeezing up around his thickness as he took them in his hands and pressed them together against his lewdly excited cock. He stroked back and forth a few times in the soft tunnel between them, flicking the tautly erect nipples with his thumbs at the same time, and then he moved up farther until his muscular young buttocks pressed down on her breasts. Lucy saw the tight bulges squeeze out as his weight sank down, smashing them onto her chest. His lustfully thick penis stood straight out from his belly, the bloated rubbery head only a few inches above the girl's gasping mouth. John reached one hand back behind him and with a quick twist of his wrist, shoved his middle finger up between her wide-open thighs, skewering it far into the hotly moist depths of her cunt all the way to the flatness of his palm. Akasma jerked as he rotated the skewering finger around inside, teasing the wet sensitive walls until finally she gasped aloud, breathing a moan of surrender to the lascivious ministrations the young bridegroom was working on her body.

His incredulous bride watched John's face twist into a obscene sadistic grin as he reached down with his free hand and pressed the underside of his rigidly swollen cock in a slow teasing circle around Akasma's open lips. Her tongue flicked hungrily out from between her teeth, licking at the tiny opening in the gland at the end. She strained forward, the cords in her neck standing out, trying to suck the tantalizingly stiff instrument into her mouth, but John laughed above her as he kept it just out of reach, allowing only the tip of her tongue to lick up and down the lewdly inciting cock's underside.

Lucy's breath was coming in quick short gasps as she involuntarily watched the depraved scene. Suddenly she could take it no more – to see her husband of one day relishing these perverted unnatural acts on another woman's body was just too much for the young innocent bride.

"Turn it off! Please, turn it off!" she pleaded, crying out as if in physical pain.

Responding to her words, Onan quietly got up and switched the set off, at the same time lighting a lamp that shone with a soft reddish glow.

"I think you'll agree that your husband has very little interest in finding you at the moment," he said, turning to stare coldly at

her confused, grief-stricken face.

"Yes, yes," the deserted young bride sobbed. "He doesn't care about me at all!" Then she completely broke down, feeling more alone and lost than ever before in her life. Suddenly without previous warning, she felt Onan's hand slide down into the embroidered robe and over her upper back, gradually forcing its way under her arm toward the rounded fullness of her breast.

"You might as well relax and enjoy yourself," he gloated, ignoring Lucy's resistance as she clamped her arm swiftly down over his probing fingers, trying to trap them before they reached her trembling breast.

"Don't!" she cried out in alarm as she realized that he expected the sight of her husband with another woman to convince her to go along with his wishes. "Just because John is shameless in giving his attentions to somebody else, don't believe for a minute that I'm just like him," she protested in a proud voice. "Because I'm not!"

With that she suddenly sat forward and struggled to her feet in order to escape his grasping fingers. But Onan had expected the movement and deftly hung onto the gown she was wearing so that it came away from her shoulders and slipped down over her arms. She pulled frantically to get away from him, but his greater strength held her leaning forward in front of him, her firmly ripe breasts nakedly exposed from the robe now lowered almost to her waist. Her arms were helplessly pinned to her sides by the sleeves of the garment, and she felt him wrap an arm tightly around her waist forcing her body back against his. His free hand slipped around her naked shoulder, sliding across her bare midsection and rising to cup her right breast. He kneaded greedily at the resilient white mound in his palm, trapping the hardened sensitive nipple between his thumb and forefinger and squeezing until a tiny excruciating sensation brought a gasp of surprised excitement from Lucy's open lips. He pushed forward with his pelvis to wedge his covered hardness into the narrow crevice between her tightly clenched buttocks.

Then the relentless slave-master crushed her ribs so tightly in his cruel embrace that she thought they would break, and he leaned back forcing her feet up off the floor. She screamed loudly and kicked her knees up in front of her as he carried her with ease over to the double bed. He held her at the edge of the four-poster until she had struggled herself into weakness, all the while running his hand from one excitedly heaving breast to the other, tweaking the pinkly erect nipples and fondling the fleshy warm mounds in appreciation. He was enjoying her vain attempts to resist, feeling her taut buttocks clenching and unclenching desperately around his hardness through the material of their robes. Finally she had no more strength to continue the struggle, and she grew limp in his powerful grasp, a warm and breathing rag doll, a play thing to be used in any way he wished. Her head hung forward with her chin on her chest, the long blonde hair falling away in cascades over her fearfully heaving breasts.

"That's right, little one, you might as well give in," he growled behind her, his breath coming in short gasps from his effort. "Nobody's going to help you now."

Without warning he suddenly threw her forward onto the bed in front of them, and without hesitation fell on top of her. As he lay there with his heaviness pinning her to the mattress, he used the freedom this gave his hands to remove the robe he was wearing, yanking it from beneath his body. Then his hands reached down to the hem of her robe that had bunched up around her knees and slipped it slowly up between their two figures, crumpling it in a tight ring around her waist. Now his distended hard cock was pressed right up into the naked warmth between her tautly squirming buttocks, its thin clear seminal fluid seeping out to make a wet trail on her creamy white ass-cheeks.

The hopeless young bride jerked and

moaned as she felt his probingly hot shaft come into contact with her naked buttocks from behind, and then she felt one of his hands trail a slow teasing path down her side and insinuate itself under her ruthlessly pinioned body to pinch at her quivering belly beneath. He forced the searching fingers between her down-grinding belly and the velvet-covered mattress until he could feel the thin curly tangle of pubic hair on the little mound at the base of her stomach. He played with it for a moment and then suddenly curled his hand down into the moist narrow slit up between her thighs, the nail of his extended middle finger scraping gently over the tiny bud of her hotly throbbing clitoris, sending chills of unwanted sensation thrilling through her body. Lucy felt a sudden surge of renewed energy to resist when his hand curled still farther into the juncture of her loins, parting the slightly moist lips of her pussy to teasingly snake its way inside the tight elastic opening of her cunt. Wild electric shocks jangled through the struggling bride's nerve-endings as she thrashed and twisted under the weight of his body above. The rigid thickness of his hotly excited cock was now pressed deeper up into the sweat-moistened crevice between her ass-cheeks, throbbing as though it had a life of its own. Lucy unconsciously ground back up against it in her panic, causing her tiny puckered anus to feel the direct contact of its bloated pulsing tip, and she flinched away from its hotness back down onto his hand.

Onan began a gentle rocking motion forward against her ass, obscenely matching the rhythm with that of his finger fucking into her warmly encasing vagina below. He could feel the exciting swell of her shivering buttocks around his hotly probing cock and the wild sensation almost caused him to shoot his seething sperm over her squirming white flesh. But he wasn't yet ready for that variation. Pressing his flat palm hard into the small of her back to keep her from raising up, he pulled his other hand roughly out from under her thrashing resisting body and tangled his fingers, wet with her vaginal juices, in her hair. Holding her fast to the bed with the pressure of his hand, the sadistic grin on his face widening, he lifted her contorted face by the hair up off the mattress. At the same time he scrambled over her on his knees until his loins were aligned next to her face, thrusting forward to ram his thick turgid cock up against her tightly compressed lips.

Lucy had her eyes clenched shut when she felt her mouth being forced open by something hard and rubbery pressing between her lips. She tightened the hinges of her jaws, but the force was relentless, and then she couldn't believe it when his lustfully swollen penis, thick and rounded at the bloated tip, slipped in between her teeth, intruding firmly until it was sunk deep in her gaping mouth. She was nearly choking on the hotly pulsing thickness, yet she pretended not to know what it was holding her teeth apart and forcing her tongue out of its way, but she couldn't ignore the insistent mouth-filling presence. She wanted to bite down hard but she was afraid her 'master' would hurt her if she did, and she thrashed with her shoulders to try to break away.

But Onan tightened his fingers in her blonde hair and shoved down harder with his other hand until it seemed he would snap her spine, his lewdly throbbing cock pushing in all the way to the very entrance to her throat. He started stroking in and out of her face, looking down to enjoy the clasping oval her unwilling lips made around his thickly rigid shaft. The hot wetness of her saliva made him groan in ecstasy above her.

"Suck it, little one. Use your tongue," he commanded her with a gasp.

The harshness of his tone and the back-bending pressure of his hand caused Lucy to open her eyes wide in fright so that she could now see the bulging thickness of his rod skewering into her face, the black mass of pubic hair actually touching the end of her nose as he pumped into her painfully filled mouth. Her neck twisted to one side was hurting so badly she thought she would faint, and secretly she hoped she would – to blot out the horror of what was happening to her.

It can't be, it can't be, she repeated over and over in her fear-clouded mind. But the incessant gagging pressure of the cock slipping in and out of her burning lips did not let up, and she responded to the tearing of his fingers in her hair by reluctantly sliding her moistly ministering tongue over the bulging vein on the underside of his shaft.

"That's it, little one," he groaned down at her, easing the grip of his fingers in her hair a little. "Suck it well, and I'll treat you well." He moved the base of his hand across the plane of her cheekbone to press down on her bulging cheek and increase the pleasure she was giving his obscenely impaling cock. She was grunting involuntarily with each inward stroke he made, increasing his sensations a hundredfold by the animal-like sounds she was making around his hotly pulsing shaft.

Experimentally, the lustfully aroused Turk eased the pressure of his hand on her lower back, and when she automatically tried to buck her body up from the bed and escape his cruel impalement of her mouth, he slapped it back down hard, bringing a low groan of helpless agony from her lips locked around his mouth-fucking cock. Again, slowly, he lifted his hand, and this time she did not make the attempt. He then brushed his now free left hand over her trembling rounded buttocks and slipped his extended middle finger lightly along the clasping crevice between them. After a moment he slipped his hand up once again to the robe that was still bunched up around her waist holding her arms like a straight jacket. And without a word, but still ramming long drives powerfully up into the hot wetness of her open mouth, he reached under her and loosened the gown to pull it away. When it caught at her bent elbows, he jerked at it until she felt it would break her arm so that she reluctantly struggled to help him get it off. Finally it came away and then he uncovered her other arm, carelessly throwing the robe onto the floor beside the bed. Now she was completely naked under his lewdly sadistic gaze, and she groaned again as he ran his fingers all over her exposed tingling flesh, pausing to tease for a moment at her tightly puckered anal ring.

"Don't think you can escape me now," he warned her in a husky lust-thickened voice. He watched her look up at his face with her terror-stricken blue eyes as if to say she had given up and whatever he wanted to do now was unavoidable, and he released his grip on her hair. When she continued to swirl her tongue in servile acceptance around the tip

of his smoothly pummelling cock, he reached down with both hands and in one swift motion turned her over onto her back. His hard throbbing shaft did not come out of her mouth as he moved over with her and straddled her face, his knees planted on either side of her head, his eyes facing the nakedly prostrate length of her lush young body. He kept stroking his expanding rod in and out, leaning forward then to grasp with his hands under her knees and raise them up and out to the side, obscenely exposing her narrow pink slit up between to his lascivious gaze. He felt the young blonde's mouth hesitate momentarily on his cock as though she might resist this further assault on her helpless body, and he immediately pressed his palms into the firm inner flesh of her thighs, spreading them farther apart and holding them there by sheer force. He dropped his head with an animal growl of hunger and began to lick at her now openly spread cunt. He could feel the softness of the fine blonde pubic hair against his lips as he flicked his tongue along the full length of her forcedly expanded vaginal opening.

Lucy jerked from the sudden electric contact with the sensitive flesh of her cunt, unexpected thrills of pleasure welling up uncontrollably inside her quivering belly. Then she felt his extended fingertips stretching the wet outer flanges away from the glistening furrow of inner softness, and his warm moist tongue drove excitedly straight through the tight elastic ring up into her startled cunt. The sudden searing joy emanating from the penetration of her vagina caused her to involuntarily bite down and scrape her teeth over the velvety skin of his rigidly throbbing cock imbedded in her mouth, and he groaned from the mixed pain and pleasure the shock sent into his cum-laden balls below.

Then Onan felt the response he had been waiting for when she slowly began to undulate her loins against his spearing, hot tongue, the tiny clenched opening of her cunt sucking hungrily at its pointed tip. He thrust as deeply as he could into the hotly expanding moistness, his lips nibbling at the palpitating outer folds of her curl-lined pussy each time he drove his skewering tongue in. He swirled his stiff oral member around and around inside the throbbing walls of Lucy's vagina, savouring her loss of control more than he had his victory over any woman before. She had been so proudly resistant up to this point that he had begun to doubt whether he could break her down without the use of some conscience-killing drug, but she was beginning to come out of her prudishness. He was conquering her by sheer will power. He knew how to turn women on.

Lucy felt like a cheap shameless whore.

Here she was spread-eagled on her back under a total stranger, his mouth doing awful obscene things down on her pulsating cunt, his cock pumping relentlessly into her mouth and, although she found it difficult to believe, she was beginning to enjoy it! She groaned heavily around the thrusting thickness of his pounding cock-shaft in humiliation and debasement – but she continued sucking him, increasing her efforts to give him pleasure with her lips and tongue as he gave her more and more of the lustful sensations she had suddenly learned to crave more than life itself.

Unconsciously the awakening captive bride began to rock her openly spread loins up and down under the Turk's face locked between her thighs, grinding her quivering buttocks lewdly down into the mattress with each down-stroke, then arching her back to reach up again for his flicking tongue. Her hips and thighs began to tremble out of control from the wildly erotic feelings he was sending through her whole body, and she thrashed and squirmed like a madwoman under him, her legs kicking up in the air toward the ceiling and jerking insanely above his head.

When she opened her eyes, she found that she was looking into the mirror over the bed. They made such a salaciously obscene picture that she was struck as if by a blow to the stomach with the lustful and, at the same time, humiliating vision. In spite of an almost overwhelming desire to shut out the scene by again closing her eyes, she stared fascinated at the dark attentive head trapped between her creamy white thighs bobbing like a man caught in the rapids of a river, and at her own face distorted into a lewd mask by the swelling thickness ramming in and out of her mouth. Glancing to the side, she found she could get another view. They formed a two-headed figure she now knew was what people called a "sixty-nine". She pulled her knees back past his shoulders to better see what he was doing that sent such wild sensations through her nerves like lightning flashes of passion, and the movement exposed her tiny puckered anus more openly to him.

The hedonistic Turk did not hesitate, but shoved his head forward the extra necessary inch to bury his tongue this time in the hot little rectal opening, bringing deep-throated moans from her chest. Then he began alternating between the two nether orifices until the thin ridge of slickly smooth flesh between them was slippery wet with the mingling of his saliva with her cuntal juices.

Onan could tell from the way her straining belly was beginning to jerk faster and faster under him that she was nearing a climax, and he took his hands from the insides of her thighs and wrapped them around her hips to grasp her lusciously rounded buttocks in his fingers, digging harshly into them with his nails as he pulled her tighter up to his mouth. He could feel her hotly quivering clitoris vibrating under his voracious sucking lips like a tiny cock about to ejaculate in his mouth, and then her cum-juices began to flood out copiously over his fucking tongue and she moaned and grunted her release around his pistoning cock. She was sucking with complete abandon now out of deep-felt gratitude for the intense pleasure this strange man was giving her, and he felt his desire-bloated balls swell and ache, ready to explode into her eagerly labouring mouth.

Her hands, which had been lying loosely on the bed beside their obscenely coupled bodies until now, suddenly seemed to take on a life of their own as they reached up to play over his pumping buttocks and instinctively massage his cum-laden balls and the base of his thick cock-shaft until the unendurably excited Turk could hold back no longer. He increased the speed and depth of his strokes into her mouth until it seemed to her she would have to choke on his hard penetrating rod – then he felt the turgid steaming cum boil up out of his testicles and begin to spurt in endless streams into her repeatedly swallowing throat. On and on came the wild gushes of thick white sperm spewing hotly up into Lucy's throat, making her Adam's apple bob up and down as she struggled to keep from choking, the burning liquid searing all the way down into her belly. Finally with one final jerking spasm, it was over – and his depleted cock began to deflate in her mouth as she continued sucking softly at the palpitating rod of flesh. He fell over on his side next to her on the bed, the two of them panting heavily from the draining effects of their orgasms. Lucy was staring blankly into the mirror above the bed, her unseeing eyes fixed on the lewdly splayed-out female body which did not seem to be her own, but that of some stranger she had never seen before.

After a long time she came to her senses, and her mind was flooded with shame and humiliation at her wanton abandon under Onan's ministrations. She groaned loudly into the room at the obscene sight of her young body stretched out openly beneath the mirror, and tearfully covered her eyes with her forearm to block it out. If only she had never married John, she was sobbing to herself, she would never have been brought to this, and she would be safe in her apartment in Philadelphia. Because of her young husband's uncaring brutal rape on their first night of marriage she had been more used and degraded than she could have imagined in her wildest nightmares. It wasn't fair; she

shouldn't have been forced to go through what she had just been subjected to!

She would never be able to forgive John for this, nor herself for being so caught up in her own physical sensations that she had actually responded to Onan's depraved caresses like one of the two-dollar whores John had gone out earlier to find in the streets of Istanbul.

I'll find a way to get even with John for this, she was thinking bitterly – but at the same time she knew she would never be able to obliterate this experience from her tortured memory. This would be with her until the day she died, and nothing she could do would enable her to forget it. Her virginal innocence of only twenty-four hours ago was gone forever.

Chapter 6

John Dean felt as though he had been transported to another planet by the intense physical pleasures Akasma had introduced him to the last hour. Now he was lying on the bed with her, totally exhausted after God knew how many orgasms and positions they had tried in making love. He couldn't decide if the overwhelming effects were predominantly from the hashish she had showed him how to smoke from the water-cooled hookah, or mainly from the wild sensations of their wantonly abandoned sex together. He finally decided it was a combination of both.

He raised himself up on one elbow and looked down at the girl he had met in the hotel lobby only a short time ago while he was looking for his wife. He wondered then why she had so willingly offered to help him and now he understood. Her lips were turned up in a sleepy sensual smile of satisfied bliss. Her pink-tipped breasts were rising and falling evenly as she rested from their exertions together.

For once his limply hanging penis did not stir as he ran his eyes over the graceful curves of her thighs and hips and the point where they melded together at the curly triangle of dark pubic hair, and on up over her taut white belly to the fully rounded half-moons of her breasts. He felt completely satiated for the first time in his life.

Suddenly he remembered how the seemingly endless orgy had begun. She had said that she had asked the cleaning woman in the hall about Lucy, and had been told that his young wife was in the hotel. John had jumped at the revelation, eager to go then and there to confront his runaway bride, but Akasma had stalled him off. She had hemmed and hawed for a long time, meanwhile preparing dark brown chunks of hashish in the water-pipe, and finally he'd got her to come out and tell him what was on her mind.

He could hardly believe his ears when she'd said that she would take him to Lucy – but only if he would make love to her first. John had been surprised at her suggestion, but he had no difficulty in complying with her request. He had been strongly attracted to her from the moment she had offered to help him in the lobby, and then to hear her beg him to fuck her led him to believe she must be a nymphomaniac. It had made the prospect even more exciting to him. Besides, he now reflected as he looked down at the sleeping girl, what choice had he had when you came right down to it? Even if she had been lying about knowing where Lucy was, that was no reason to refuse, especially since he was already interested.

Now he no longer felt any passion raging in his loins, however, and he reached over and shook Akasma by the arm. Her eyes opened immediately as though she had been watching him secretly from beneath her half-closed, sultry eyelids, and she smiled up at him seductively.

"Again?" she asked, mistaking his reason for waking her up. "You Americans have a lot of stamina."

"Not this time," he said, "I want you to

take me to my wife – now," and he squeezed her upper arm painfully in case she wanted to stall him again. He had had about as much as he could take of her demanding sexual tastes, at least for the moment.

Akasma looked up at him speculatively, as though trying to decide how much farther she could push him before taking him to the next phase of Onan's plan. She couldn't be sure how he would react when he learned what his wife had been subjected to, and she would have liked to enjoy a little more of his youth and physical strength. Then with a sigh she made up her mind that he wouldn't wait any longer, and she slowly got up from the mattress, giving his deflated cock a final playful squeeze with her fingers before standing up.

"All right," she reluctantly agreed. "Get dressed and we'll go. You can put on that gown over there; we won't be going far and no one will see you."

He shrugged into the fancy silk gown she had pointed out to him and waited for her to straighten her hair and join him at the door of the bedroom. She lifted her face up to his for a quick kiss before leaving, and he felt her tongue dart serpent-like into his mouth, but he ignored the light aching sensation this caused in his depleted testicles as he urged her toward the door.

"Promise me you won't say anything until we find out if she's still busy," she said as they walked together down a carpeted hall from her apartment.

John's heart jumped in his chest at her words. "W-what do you mean, busy?" he demanded.

"You'll see," she smiled up at him coyly. "Just promise me, or I can't take you there." She stopped and refused to go on until she had his word.

"Oh, all right, all right," he answered in exasperation. It was obvious there was something going on he didn't yet understand, but he would never find out if he didn't go along with her. Strange thoughts of what he might see were running through his mind as he meekly followed her to another old section of the hotel.

Finally they stopped in front of a tall door of heavy carved wood, and Akasma pulled a silk tasseled cord. In the interior, as if from a great distance, John heard a softly tinkling bell, and the door was opened by a huge man with a bald head and muscled arms as thick as the average man's legs. He stepped aside after shooting an inquisitive glance at John, nodding solemnly at Akasma as though he knew her and beckoning them to enter.

What the hell has Lucy gotten us into? John asked himself as he followed the Turkish woman from one spacious room to another. Finally she put up her hand to signal that he should not talk, and she slowly opened another door and leaned her head inside. She watched whatever was going on for a long moment and then she turned back to John, whose curiosity was getting the better of him, and leaned close to whisper in his ear.

"Remember, you promised not to say anything until they're finished," she warned him, looking deep into his eyes to see if he would do as she said.

Finally satisfied that he understood, she opened the door a little wider and slipped silently into the room. John followed close behind, his heart beating in his chest like a tom-tom. He couldn't imagine what Akasma's mysterious actions were leading up to.

And then he saw Lucy!

His lovely young bride was up on her knees in the middle of a huge bed surrounded by mirrors, her face pressed down on its side into the mattress, her young body completely naked. Up behind her openly exposed buttocks a man John had never seen before was kneeling, staring down at her defenceless genitals obscenely proffered to his lustful gaze. The man held his rigidly erect cock in his right hand, preparing to insert it between her lewdly spread ass-cheeks where the young beauty's husband could see the pink moist slit of her vagina below.

John gasped in disbelief. This couldn't be his innocent inexperienced wife waiting expectantly on her knees for the long thick rod poised inches from her upraised buttocks. Not the naïve, reluctant little bride he had so recently fucked for the first time in her life. No... it just couldn't be!

Then the girl lifted her face from the bed a few inches as if to look back at the man behind her to see why he was hesitating, and John hissed out his breath. He no longer had any room for doubt that it was in fact his bride. His stomach muscles tightened and he was about to cry out in anger when he felt Akasma's hand tighten on his wrist as she gave him a threatening glance. He stopped, fighting down his sense of outrage at the scene that was taking place only a few yards away on the huge bed.

To all appearances, neither Onan nor John's wife Lucy had noticed them enter the room. The captive blonde was suddenly quivering in fear as she felt the intimidating hugeness of the blood-swollen penis pressed into the split between her widespread buttocks. He had ordered her up on her knees in a harsh voice that she could not ignore, and, thinking that she had nothing more to lose after the depraved sucking act he had just forced her into, she had obeyed his command.

Oh God, he's too big, she was now thinking in panic. He'll split me open!

Her buttocks involuntarily cringed forward drawing away from the rubbery tipped cock, but the relentless shaft followed. Lucy's forehead was once again pressed tightly down into the mattress. There was no way to escape his hands grasping painfully into the firmly flaring flesh of her upraised half-moon buttocks.

Lucy's incredulous husband saw her cringing forward, and, after a first impulse to rush to her aid, he smiled crookedly to himself. Let him tear the phoney little bitch open, he swore to himself. After the way she had resisted her own husband scarcely one day before, he thought bitterly, she deserved what she was about to get.

On the bed the frightened young wife felt Onan's hands slip forward and close around the curving swell of her hips, holding her firmly back against his hardness.

"Reach behind – put it in," he panted in anticipation of what was coming.

"Oh, no, I can't," Lucy whimpered into the bedspread. "I just can't; it'll hurt!"

"Put it in, I said," he growled down at hers digging his fingers still harder into her flesh.

"Ohh," Lucy groaned as she felt the upraised flesh of her hips being squeezed into tight little balls. She couldn't stand it.

In desperation she reached back between her legs and closed her hand over his penis. The excitedly throbbing cock-shaft felt even larger than before when the masterful Turk had fucked into her mouth. She would never be able to take it.

John stood with Akasma near the door watching the scene. He felt his cock jerk slightly from the lewdly inciting picture in front of him as he saw Lucy's hand close around the dark-haired man's thick penile shaft, holding it poised barely an inch from the up-thrust entrance into her proffered, cunt that looked so vulnerable. The exciting tableau was reflected from various viewpoints in the surrounding mirrors and the young husband felt a sinking sensation of jealousy and desire mingle in his stomach as he groaned softly under his breath.

Onan squeezed her defenceless hips again, this time bringing a louder squeal of pain from Lucy's tormented lips. She tearfully placed the rubbery throbbing cock-head against the clenched elastic opening of her vagina, biting down hard on her lower lip to hold back the tears of fright that were brimming in her eyes. She felt the passion-bloated glans begin a slow prodding against the slightly open lips of her trembling pussy, parting them and forcing its way inside the tight restricting ring of flesh that protected the entrance to her secret inner passage. There was a great stretching feeling in her loins as though the tender lips were being pulled asunder, and then suddenly she felt her velvety cuntal ridges swept apart and Onan's hard thick cock slip up into her hotly throbbing vagina like a blunt pole of flesh.

"Auggggh!" she choked out through clenched teeth. Her vaginal walls went into wild, fiery spasm. The lust-swollen penis penetrating her fearfully squirming cunt felt like a thick drill tunnelling deep into her belly through the involuntarily contracting inner passage. She struggled and swung her buttocks from side to side in a vain attempt to escape the ruthless impalement. It was no use. Onan pushed forward relentlessly, skewering her right up to the hilt, making her a hopeless prisoner of his rigidly pulsating cock like a beautiful, fluttering butterfly pierced through cruelly by a pin. Her lips opened and closed in pain and torment as she moaned into the mattress from the outward pressure of his deeply imbedded cock-shaft. He slipped it back out a fraction of an inch and then groaned huskily as he shoved it forward again up into her pussy depths as far as it would go. She felt its pounding rubbery tip bump up against her cervix at the same time as his heavily bloated testicles slapped up hard against the softness of her pubic mound below.

The impaled bride's husband watched,

his eyes wide-open, as the fat stocky rod disappeared up inside Lucy's tight, widely stretched vagina. John moaned softly in unconscious vicarious pleasure as he imagined the hot moistness clasping around the stranger's cock buried deep up inside his young wife. His own penis so recently satiated with fucking Akasma jerked wildly inside the loose-fitting robe, pointing out from his belly, its swollen head popping obscenely out through the opening in the front of the garment.

The lovely Turkish girl couldn't help but notice his hardness thrusting stiffly out into the room, and, with a tiny mewl of appreciation, she grasped it in her hand, slowly massaging the loose outer skin up and down over the throbbing gland at the end. She reached with her free hand down inside her own robe and slipped her extended middle finger up into the moistening slit of her pussy, squeezing her excitedly quaking thighs tightly around her hand to heighten the excruciating sensations of pleasure flooding once more into her loins. She kept her lust- and hashish-dilated eyes fixed avidly on the lewd scene on the bed, flicking her finger tip lightly over her erect clitoris in time to the steady in and out strokes her handsome master was beginning to make. She wondered if Lucy would experience the same degrading anal invasion that she had herself only the other day.

The helplessly skewered young blonde heard Onan behind her, gasping and panting lasciviously as he began to rock rhythmically back and forth into the wetly responding confines of her expanding vagina. Gradually the pain of his entry was subsiding, giving way to a feeling of strange masochistic pleasure that rippled through her, sending waves of sensation through her insides that cried out for more of the same. She began to slowly undulate her buttocks in erotic little circles back onto the hotly spearing shaft fucking into her from behind.

"Ohhhh, God," Lucy suddenly breathed back at him in a tremulous voice. "I like it, it's good. Keep on, keep on!"

Her astonished husband watching from the doorway couldn't believe the thick husky sound of lust coming from her lips. She didn't seem to be the same woman any longer, and he revelled in the cruel humiliation she was being subjected to. It served her right for depriving him of this kind of passionate response the one time they had fucked.

Kneeling up behind Lucy's bent-over body, Onan gritted his teeth and fucked in and out with long leisurely strokes, waiting to feel the signal of her total acceptance inside her responsively swivelling cunt. Her words had shown him she was beginning to like it, but he knew, if he took his time and worked patiently, she would soon be begging him for more and more. He had been anticipating her complete surrender, and he wanted it like this, not under the influence of whatever drug it was that Hasad had given her at the motel. He wanted to know it was her fully conscious mind that was giving in, not a mindless robot of a woman who didn't know what she was doing. A feeling of absolute power surged through him as he held her hips in his hands and skewered repeatedly into her, enjoying the sensation of the soft fleshy ridges deep inside her cunt slowly opening around his insistently pistoning cock, her lust-juices hot and slippery in the tight narrow passage.

He stretched the quivering half-moons of her buttocks wide with his fingers, watching with delight the pinkly wet, glistening cuntal walls clinging around his penis on the out-stroke, then folding in with the impaling instrument as he drove repeatedly back into the snug vaginal sheath. He sensed that she had never been fucked this deep before and he rotated his swollen cock-shaft around inside, enlarging the moistened passageway and feeling it clasp greedily around his pulsing hardness like warm elastic. The soft enveloping tightness brought a tingle of maddening delight shooting through his cum-swollen balls as his wetly encased cock pulsated with lewd pleasure.

Lucy's eyes opened and closed in a smoky glaze of passion, flashing like blue signal lights from the intensity of the feelings ravaging her plundered loins. She spread her thighs out wider, moving her feet apart on either side of Onan's bent knees. Her head dropped lower onto the mattress, forcing her swaying white buttocks up higher for greater contact, and her heaving breasts pressed down flat on the bedspread, the rough material tantalizing her tautly erect nipples. Forgotten now was the humiliation of bending like a slave while a tyrannical stranger fucked her relentlessly from behind, forgotten was her puritanical aversion to promiscuous sex. All that mattered at this moment was the obscene pleasure coursing through her like a fire raging out of control. She wanted to be fucked like this, she wanted to be subjected to any lewd demands placed on her passion-wracked body.

Looking down from his position above her, Onan saw the small brown puckering anus nestled in the narrow crevice of her openly spread buttocks, and suddenly he wanted nothing more than to sink his cock into that virginal untried opening and cum inside her rectum. He slipped a finger of his right hand down into the wetly flowing juices seeping out around the pink flanges of her pussy, moistening its tip thoroughly before sliding it up again to the tiny anal mouth clenching and unclenching under his gaze. Lubricated by her cuntal fluids, the finger paused momentarily to tickle lightly at the puckered entrance and then without warning he shoved it forward, sinking it in up to his fist.

"Arrrrrghh!" the startled young blonde groaned in front of him, her feet jerking up and down on the bed as she vainly struggled to escape his brutally impaling finger. Lucy lovely features flushed red with both discomfort and embarrassment. He can't... put it in there, she thought naïvely, that's... that's where I go to the *bathroom!* But he gripped her shoulders forcefully with his free hand, pulling her back tighter against him. Slowly he swirled his finger around inside the narrow contracting passage, working excitedly to expand and enlarge it for the greater thickness to come.

Lucy felt the discomfort in her anus slowly subside as her rectal passage grew accustomed to the presence of his skewering finger, and she gradually began to rock back to meet his inward thrusts, matching his steady rhythm stroke for stroke.

Her husband John couldn't believe his eyes when suddenly, without warning, Onan pulled his finger out of the now expanded anus with a soft popping sound, and, in one swift movement, changed his cock from one opening to the other.

On the out-stroke his powerfully pistoning penis had been in her swiveling cunt, and now on the in-stroke it was shoving its way by sheer force up into the narrow confines of her helpless little anus. He tightened his grip on her shoulder muscles with both hands now to hold her in front of him, relentlessly shafting up into her anal orifice, rushing up the tight canal until his stomach was pressed flat against her wildly jerking buttocks.

"Agggh, Augggh, Uuuuuuugh! You're putting it in my *bottom! Please...* not there – it's too big and... it's so... *dirty!*" Her voice trailed off and Lucy groaned under him from the sudden tearing entrance into her unprotected rectal passage. This was not merely uncomfortable as when he had pushed his finger up her bottom. This was *agony!* She thrashed and bucked helplessly before him, her eyes bulging open and her long blonde hair whipping from side to side as she tried to throw herself off his excruciating thickness. But he had her fastened inescapably to him now, and she could feel his penis throbbing lewdly inside her like a heart beating in her rectum, each pulse shoving her cruelly stretched inner walls a little farther outward from his hardness, until finally she could breathe again. Her body felt almost limp to her tormentor, like a rag doll down at his loins, as he began to stroke smoothly in and out, short and shallow motions at first, but then gradually lengthening until just the expanded glans at the end of his excitedly pulsing cock-shaft was left inside before he shoved it back in her right up to the hilt. To the obscenely skewered young bride, it felt as though his cock was gliding wetly all the way up between her labouring lungs each time he plunged into her. She moaned and perspired under him, not yet moving in response, for although the pain had lessened to extreme discomfort, it was still too much for her.

Then the strangest thing happened. With a low mewl, like an animal in heat, the beautiful blonde began to rock gently with him, meeting his inward strokes with her back-thrusting buttocks, grinding the white half-moons into his hard stomach lasciviously to get as much of his lust-liberating cock as she could. She was enjoying it now, and her mouth hung slackly open to the side as she gasped and panted out her passionate pleas in all the inciting obscenities she had ever heard or read.

"Ohhhhhhh," she fluted in a high, breathless voice. "Fuck me, fuck my ass! Ohhhhhh, fuck my ass!"

The sadistic Turk gritted his teeth from the tightness of the narrow hot anal ring of muscle, pulling his lips back over his teeth in a grimace of mingled pain and pleasure. He had conquered her as completely as possible now, and he laughed between his clenched teeth in low grunt-like sounds as he began pounding harder and harder against her creamy flattening buttocks.

"Unh, unh, unh, unh," she grunted in time with his smooth strokes up between her ass-cheeks, the sounds all but smothered out by the saliva-soaked patch of bedspread bunching up in front of her mouth. Now the pain she had felt at first was replaced by a different, aching pain of unsatisfied desire, and she arched up her body to let a hand pass under it, slipping up over her trembling belly to burrow into the wetly expanded furrow of her pussy. Her fingers found the quivering clitoris and massaged frantically at the swollen bud to try and quench the hunger inside her hotly throbbing cunt. She could feel Onan's heavy balls banging up into contact with her hand with each thrust and she grasped their hairy softness and began to milk them for the hot orgasmic cum building up inside.

She was approaching orgasm and her body had become something animal, no longer human as she twisted and churned under him, jerking her buttocks back to goad him on like a wild stallion riding her in the race to completion.

"Goddamn," she heard a strangely familiar voice say through her lust-hazed mind. "Look

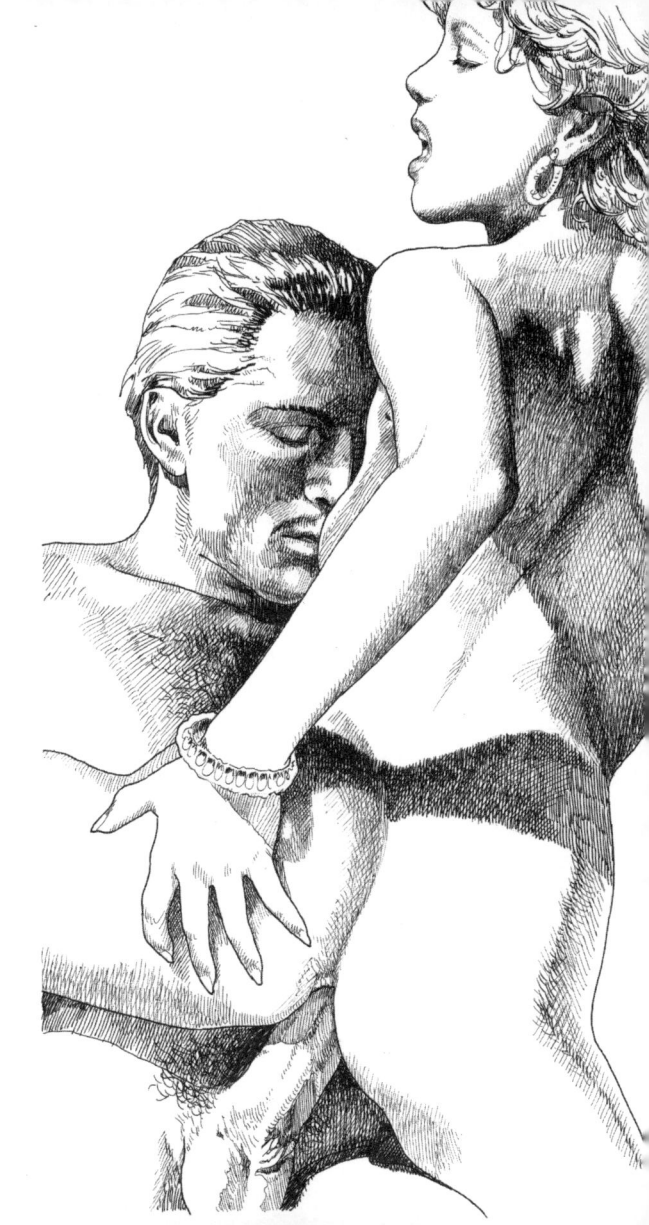

at the little bitch take it up her shit-chute!"

Lucy jerked her head up from the bed in alarm. There was no mistaking that hissing voice now – it was her husband John. She dimly saw him in the distance watching her humiliating sexual abandon, his stiffened penis sticking excitedly out in front of him through the robe fallen open in front. A woman stood close beside him, her hand stroking the outer skin of his lustfully hardened cock up and down as she finger-fucked herself with her other hand.

Oh, God, no! the shocked young bride screamed inside her confused mind. It's John, and he's standing there watching as though he doesn't even know who I am! Then she remembered where she had seen the woman before – on the television screen making love to her husband – and his young wife looked directly at John with her lust-crazed eyes, smiling a lewd seductive smile to torment him with her wantonness.

John was suddenly filled with a vengeful anger that this was happening to his own wife, after she had been unable to satisfy him only hours before. He swept away Akasma's hand and ripped off his gown to stand completely naked in the room.

"Ram it in her, friend – whoever you are. Give her all you've got!" he growled.

At that Onan finally took notice of what

was going on behind him, looking into the mirror to see the young girl's husband. He hesitated a moment, breaking his rhythm in the tight little anus, and then suddenly had an idea. He would show both these superior-minded Americans the depths he could drag them to, he thought with sarcasm.

Locking both his arms around Lucy's wasp-like waist, he pinned her back against his loins and with a sudden rolling motion turned over onto his back. She was still impaled on his thick hard cock, but now her dilated cunt was nakedly exposed to the others in the room, her legs wide-flung to the sides as she struggled to find a foothold to balance herself. The muscles of her flat white belly were straining to the breaking point from the pressure of her legs falling down onto the mattress on either side of his body under hers, and finally she managed to plant her feet high up on his out-stretched thighs, her knees bent to the sides. This relieved her stretching muscles and she pressured up and down on his once more plunging cock, using his muscular thighs as a spring-board to push her buttocks up and down against him.

John could see the thick dark shaft protruding from her asshole underneath, and looked with greedy eyes at her pinkly moist cunt up between her thighs. He watched the open, hair-lined slit making tiny circles in front of him and saw her fully ripe breasts dancing on her chest. She looked so helpless and enticing that he could stand it no longer.

Onan saw the sudden heat of passion enter the young girl's husband's face over her shoulder, and he grinned in satisfied glee.

"Come on, Johnny, old chap!" he goaded him. "Join the fun. Your wife really is a fantastic fuck, old man. Why don't you fuck her cunt while I'm fucking her asshole?"

John groaned at the insulting tone and in the man's voice and he was outraged by the familiar tone he was adopting, but under these circumstances he couldn't resist. In a few quick strides he reached the edge of the bed where they were screwing and making a low-pitched growl of lust deep in his throat, threw himself on top of the churning mass of flesh. He crawled up over Lucy's wildly thrashing body, and positioning himself up between her upraised thighs, he lodged the swollen head of his throbbing cock in the mobile, twisting mouth of her pussy. Then he balanced himself with both hands spread out on either side of the two grimacing, lust-contorted faces below him, and with a sudden forward thrust of his loins, pushed up into the hotly expanded warmth of his new bride's open cunt. There was no resistance left inside the hot clasping walls, and his wildly rampaging penis raced headlong up into the very depths of his wife's soft white belly until he felt it ram against her cervix.

Lucy groaned at the renewed pressure filling her hungry little vagina, and snaked her legs up around her enraged husband's waist. She locked her heels behind his already hard-driving buttocks, pounding into them to spur him on. She had never been so utterly filled in her life, and she wished she could go on for an eternity like this, feeling the two rampaging cocks rubbing against each other through the thin wall of flesh separating the two nether entrances into her body. Lucy looked deep into her husband's eyes, her expression an extraordinary mixture of humiliation and desire. By contrast, his features betrayed no emotions whatsoever; his face was expressionless. She curled her left arm around John's neck and held onto him for dear life, at the same time pressing her other hand down between their grinding bellies until she could feel both hard, wetly slipping cocks entering her below. With her fingers she clasped at the two men's testicles and tickled them lewdly, not knowing which belonged to whom, and not caring in her totally abandoned state.

Akasma groaned out loud in the room at the obscenely exciting trio on the bed, and then she too stripped down to her naked flesh and flung herself down beside them on the mattress to watch from a closer viewpoint. The lovely Turkish girl lay on her stomach, her face only inches from the thick cocks thrusting again and again into Lucy's anus and vagina, her own buttocks waving wantonly in the air as her hands massaged her throbbing hot pussy in sympathetic enjoyment of the double ravishment taking place before her very eyes. She could see the tight ridges of elastic anal flesh that sucked at Onan's penis as he pulled it from deep within the young blonde's rectum; it was as if her body was trying to hold the big phallus captive in her bowels; she saw too the pinkly moist outer folds of Lucy's cunt clutching wetly at John's rock-hard cock-shaft fucking into her.

The recently virginal young wife shuddered ecstatically and bucked her body crazily, her eyes staring up into the mirror at the wildly twisting heap of bodies on the bed, her mind torn between the hot desire scorching her twice-impaled belly and the quick twinge of shame and humiliation that flickered through her. Her mouth was wide open as she gasped and panted for breath, tears of mingled joy and debasement streaming down the distorted planes of her cheeks.

She had been so close before to her orgasm that John's thickly expanding cock pounding into her wide-open, defenceless cunt finished it for her, the cords in her neck suddenly tightening like steel cables.

"I-I'm cumming! Oh God, I'mmmmm cummmmmming! Fuck me harder you lovely bastards! Fuck harder!" she pleaded in a voice shrill with passion. "I'm cummmming! Ahhhhhhhh!" Her voice changed pitch then, rising to a banshee wail that went on and on, broken rhythmically by the breath-taking plunges the two lusty men were making into her body. Her hand milked the tight, cum-laden balls below, pulling to hold the thick lengths of cock as far inside her as she could. Her cum juice flooded wetly out around John's still driving cock, flowing down the crevice of her buttocks to soak Onan's penis with the hot fluids that streamed on and on out of her throbbing cunt.

Spurred on by Lucy's twisting body below, John shoved his lustfully aching rod deep into his wife's cunt, fucking into her like a pile driver gone wild, and then he felt the hot waves of his surging sperm shooting far up into her dilating vagina, filling her belly to the bursting point. The mind-shaking forces of his orgasm caused him to groan as though his very life were being drained from his body, and then he held his pulsing shaft all the way up in her gratefully sucking cunt for the final spurts of white-hot semen. He was bounced and tossed wildly on top of her from the force of Onan's bucking below, increasing his pleasure a thousand-fold. He had never felt anything like it in his life, and he held onto his young bride's jerking body as though his very life depended on not falling off.

Onan no longer felt the need to wait. Quickening his tempo and arching his back until he was fucking up into her hungrily constricting anus in huge, powerful thrusts, he growled deep in his throat and felt the steaming hot cum fluids begin boiling up out of his warm, aching balls. His hands slipped between the bodies on top of his to grasp greedily at Lucy's swollen breasts, jerking her down onto him tighter and tighter as his thick, fiery cum spurted deep up into her desperately contracting anus. His exploding cock jerked like a machine-gun, the lewd sperm spurting endlessly to fill her rectum all the way to the puckered outer ring of tightly locked flesh. Finally he gave one last body-slapping shove and held her down tight on his still rhythmically pulsing cock, savouring the spastic tremors of her ass-cheek muscles around it, until he gradually relaxed the tautness in his back and let his weight fall back down onto the bed. His slowly deflating cock slipped out of her temporarily expanded anal passage with a wet sucking noise, and he released his breath in a long, hoarse sigh as if he had been holding it in a long time.

Akasma watched Onan's limp battered

penis followed by thin white streams of semen slipping out of the young American girl's slackened anus, so close she could see the pulsing veins on John's cock as it slid uselessly from Lucy's cunt down between her wide-flung thighs. Impulsively, the dark-haired Turkish woman glued her mouth to Lucy's battered genitals, running her tongue down the grove of her swollen, sperm-drenched labia until she arrived at the gaping, leaking anal orifice. Lucy bucked and heaved with this new, tender assault on her sensitive cunt and asshole, her next orgasm like the aftershocks of an earthquake, not nearly as strong as the main event, yet still intense and pleasurable; the Turkish beauty continued to lap at the river of white sperm until she had greedily swallowed as much as she could. Now Akasma turned her attention to the men: first she took one then the other cock into her mouth, briefly sucking and savouring the mingled juices and flavours. This perverse act was enough to tip her into her own wild orgasm: she felt that sudden hot broiling feeling in the depths of her belly that signalled its advent, and she pumped her ass-cheeks in the air as she ground down on her fingers buried up in her flooding cunt. She moaned and chanted obscenities in her own language, arching her neck up off the mattress to continue licking at the lewdly exposed loins inches from her face.

She shivered spasmodically for a moment, her long legs stretching out and quivering, her half-moon buttocks tightly clenched into little white balls of flesh, and then, with a gasp, collapsed into a heap of satiated flesh beside them, her head resting on John's thigh.

They lay like that a long time, their breathing heavy but gradually slowing down to normal, all their minds thinking the same thoughts.

Never had any of them experienced such uncontained joy in their lives. Not even Onan, who had spent his entire youth searching for just such pleasures. And not Akasma, who had believed she knew all that there was to know about the secrets of the flesh. John had never in his wildest dreams thought it possible to feel as completely drained and fulfilled as he did now.

And Lucy: Lucy could understand now what had possessed so many women in the past to become whores or courtesans. This passionate outpouring of physical emotion had to be the greatest goal anyone could ever hope for. Never again would she be able to live without it, even though her deeply ingrained conscience might torture her relentlessly, as it was even now – she was committed to seek physical excitement wherever she could find it.

And it was all John's fault, she thought ironically. If he had been more patient with

her that first time, she might never have learned what real sensual pleasure was like. She gave her husband's neck a tight affectionate squeeze and smiled a secret smile of satisfaction up at herself in the mirror.

Chapter 7

The late afternoon sun filtered through the gauze-like curtains covering the windows, and Lucy's eyes fluttered once and then opened wide. She was refreshed from the nap she had taken to prepare herself for the evening ahead and she stretched languidly, enjoying the feline look of her young body in the mirrors around her.

The spring of her wedding to John had passed into fall, and now it was almost wintertime, but the Istanbul sun was still strong and warm. She had spent a lot of time sunning and swimming naked in Onan's private pool, in the pleasant shaded garden behind his *yali* and her body was no longer the pure innocent white it had been such a short time ago. It was faintly apparent also that she had matured more than she should have in the last few months – tiny crow's feet were becoming barely visible at the corners of her eyes. Her makeup was a little heavier these days too, but one who had not known her before would not have noticed her worldliness. She still retained the fresh sparkle in her light blue eyes and the youthful eagerness in her body. In fact, she could feel that eagerness coming to the forefront of her mind as she waited for John to come to pick her up for the latest project they were working on for Onan.

The fast-maturing blonde had actually enjoyed the past few months more than she would have believed possible. She had been with many different men, often two or three at the same time, and she seemed to be more and more demanding every day. It was getting to the point now where she couldn't be satisfied unless she had a constant diet of lovers, and if it hadn't been for Onan's continuing attentions, she felt she would have gone out of her mind. Poor John just couldn't stand the pace she required any more, especially since he spent so much time seducing other women for Onan's purposes as well as his own pleasure. She didn't really blame him for his inability to meet her needs: probably no one man could have.

She and John had a good deal going with Onan, and they had done well over the past few months. She smiled to herself at the ease with which they had made the arrangement that first week in Onan's mansion, and she mentally went over the secret bank account they had been building up with commissions from Onan. At first she had been deeply resentful of John for the life he had led her into, but like many young Americans abroad, the way of life was so appealing in its mystery and exoticism and so liberating in its lack of puritanical constraints, that she soon lost any notions of guilt or self-reproach for leading these young girls astray – and quite possibly into a life of sexual slavery. Besides, she thought, most of them could have escaped at one stage or another – if they'd really wanted to.

* * *

Knotting his tie loosely around his neck, John Dean looked at the svelte form of his blonde, naked wife as she stood in front of the mirror to apply her makeup. He felt his cock give a familiar lurch in his pants at the sight of her pink, tempting ass: the luscious globes of her buttocks in an eternally welcoming symmetry and, just below, where her thighs joined her torso, the downy split of her vulva was just discernable. He imagined that she would be already wet there at the thought of their strange mission tonight with its equal measures of raw, sexual excitement and physical danger. God, he'd like to take her right here and now, as she leaned over the vanity unit.

"Better get some clothes on in a hurry," he told her in a thick voice. "Or I just might have to fuck you myself before we go. And I'll need all my energy tonight if things go according to plan." He reached out and playfully tweaked her pinkly erect nipples, sending shivers of pleasure rippling over her belly.

"That's not such a bad idea, lover," she replied in a coquettish tone. "But you know that it's strictly against the rules."

Onan had made it clear to them that they were perfectly free to do anything they wanted, but on days when they had an assignment for him, they had to conserve all their passion for the object of his designs.

This time it was the daughter of a wealthy Englishman. Onan had had his eye on the girl for a long time, and he had received word that she was on the verge of starting her first affair. She had gone out of the city ahead of her potential lover to wait for him at the same motel on the Dardanelles where Lucy had gone that night which seemed so many eons ago. Onan had arranged for the girl's young boy friend to be held up on business in Istanbul until Lucy and John could have a chance to get to her. Onan wasn't afraid of any repercussions if he succeeded in obtaining the girl – he could have kidnapped the daughter of the Queen of England if he wanted to. His millions protected him better than the whole Turkish army could. Besides, he had told them in the planning talk the night before, he controlled practically every source of income the English girl's father depended upon, and if it came right down to it, he could threaten him with total ruin. But he didn't think it would be necessary; he had too many other possibilities for covering himself.

However, this was one foray that Lucy didn't look forward to with much relish. Onan had told her that she would have to somehow distract Hasad long enough for John to get to the girl, and Lucy remembered the revolting clerk at the motel only too well. She hurriedly gulped down her glass of whitish raki and stood up before she could think about him too much and change her mind.

John watched her as she clothed her body with silk underwear and a mini-dress that did little to cover her physical charms, and in a few minutes she was ready to go. She thought it was a waste to wear such expensive clothing for the night manager's benefit: what she was wearing cost more than her whole bridal costume had a few months back, but she owned nothing now except the very best.

"All right, Romeo," she said with a short laugh, her sensual lips turning up in a practiced little seductive smile. "Let's hit the road."

* * *

Whistling softly to himself in the motel office, Hasad put on his blazer to go on duty for the evening. He had been watching the young English girl for two nights now, and he could tell she was nervously waiting for some man to show up. He wasn't going to wait any longer. Tonight was the night, and he had carefully prepared the way both previous evenings by solicitously offering her one of his special drinks – without his special blend of drugs. Tonight, however, would be different. He would put in more brandy than usual so that she wouldn't notice the change in taste, and then... he licked his lips greedily at the thought. Perhaps he would telephone one of his new contacts in Istanbul and make arrangements to sell them the girl when he was finished, but he would see about that. First he wanted to be sure of getting what he wanted. He still hadn't forgotten his mistake with Onan and Madame Cari months ago when the American girl had been there.

He removed from his street clothes the tiny chunk of oily hashish he had bought that morning from a gypsy, and put it into his blazer pocket. In his mind he was running over the past two nights when he had secretly watched the English girl in her bath. Luckily for him, she was one of those who liked to read in bed every night, giving him ample time to enjoy looking at her sensuous body lounging on the bed. Twice he had masturbated from the exciting picture she made lying with her legs splayed out while she concentrated on her reading. But tonight he wouldn't have to waste his precious load of sperm. He would shoot it up into her smooth white belly until she was filled to brimming with the hot juices of his lust. He felt the eager tickling begin down in his testicles and moaned softly as he smiled in anticipation.

A little later, after he had said goodnight to his fellow workers and was now completely alone in the office, he was about to make the drink for the girl, when he noticed the lights of a car entering the drive. He thought it had stopped and was going to turn around to head back in the other direction on the highway, but then it moved forward again and he cursed harshly under his breath as he went out to meet it.

"Have you got a vacancy?" Lucy asked him in a flirting tone as he opened the door of the taxi. She almost laughed out loud at the way his mouth fell open when he recognized her. She would have no trouble at all keeping him occupied, she thought, noticing the way his eyes bored into the vee between her thighs when she slid across the car seat to get out.

"Of course, Madame," Hasad answered, falling almost immediately back into his obsequious manner. He couldn't imagine how this little bitch had got out of Onan's hands, but he was secretly glad she had. And this time, he swore to himself, we'll see if you can get out of being screwed within an inch of your life. He knew the same method of drugging her drink wouldn't work, but he would think of something.

With her new understanding of men, Lucy could almost see the thoughts running through his mind, and this time she pretended not to be so aloof. She took his arm when he offered to help her out of the taxi, and she smiled with a practiced coy innocence as she thanked him.

"I don't suppose I could have number fourteen," she asked when they were inside, watching the reaction on his face.

"I'm sorry, Madame," he replied, pretending to look it up in the register. "That cabin is occupied at the moment."

Lucy had been positive that the English girl would be in that room, especially after hearing about the peephole from Onan, and she knew that by now John would be knocking at her door. He had jumped quickly from the taxi driven by one of Onan's employees before it had driven on down to the office and would have no trouble finding the proper cabin.

"Then another one will do," she told Hasad. "And could you bring me one of your special drinks? I really enjoyed it the last time I was here."

Hasad couldn't believe his ears. She was actually asking for it! The thought crossed his mind that there was something strange about her being here, even more so about her asking for the drink, but the look of open innocence on her pretty young face disarmed him. She must have remained ignorant of his part in what happened before, and he slowly let out his breath in relief. As far as she knew, he was just the night manager in a motel she had once stayed in, and she trusted him.

"Of course, Madame. It will be no trouble at all, although the bar is closed. I can mix you one here in the office. If you will follow me to your room?" he suggested, taking a key from the rack behind the desk.

"Oh, if you don't mind, I would prefer having the drink here first," she told him with a slightly suggestive smile. "I'm not expecting my husband until morning, and it's so lonely in these isolated cabins. Do you have a lounge where we could sit and have a drink together?" She smiled up at him in a seductive way that left no room for doubt what she was after.

Hasad licked his lips speculatively. He would have preferred taking her to her room